EYEING THE RED STORM

EYEING THE RED STORM

Eisenhower and the First Attempt
to Build a Spy Satellite

ROBERT M. DIENESCH

University of Nebraska Press
LINCOLN & LONDON

∞

Library of Congress
Cataloging-in-Publication Data
Dienesch, Robert M.
Eyeing the red storm: Eisenhower and the first attempt
to build a spy satellite / Robert M. Dienesch.
pages   cm
Includes bibliographical references and index.
ISBN 978-0-8032-5572-2 (cloth: alk. paper)
ISBN 978-0-8032-8675-7 (epub)
ISBN 978-0-8032-8676-4 (mobi)
ISBN 978-0-8032-8677-1 (pdf)
1. Space surveillance—United States—History—20th century.
2. Artificial satellites—United States—History—20th century.
3. Astronautics, Military—United States—History—20th century.
4. Military surveillance—United States—History—20th century.
5. Eisenhower, Dwight D. (Dwight David), 1890–1969. 6. Cold War—History.
I. Title. II. Title: Eisenhower and the first attempt to build a spy satellite.
UG1523.D54 2016
358'.88—dc23
2015028635

Set in Minion Pro by L. Auten.

I dedicate this book to four people who
have meant the world to me.

It is dedicated to the memory of my grandparents:
Michael and Susanna Dienesch and Christa
Laudenbach; unfortunately they did not live to see
the completion of this work. I never knew if they really
understood what I was working on; I do know that
they loved me and that my world is a
little less without them.

The fourth person is my wife, Jennifer Mattinson.
Before you came into my life, I was lost and without
direction. You have brought back the light and
magic in my life. I dedicate this book to you,
my cheering squad, my rock, my love.

CONTENTS

ACKNOWLEDGMENTS

Inevitably during the research and writing of a work such as this, a large number of people have given me assistance. It is impossible to give credit to each of them in turn, so I beg their forgiveness for not including a detailed list. However, certain individuals and organizations do stand out for special thanks.

As with every research project a great deal of archive work is essential. I have been lucky that the archival staff I worked with were exceptional. The knowledge of the archivists at the Dwight D. Eisenhower Archive in Abilene, Kansas, and their willingness to assist made my time there not just productive but a splendid experience. I look forward to having the opportunity to work with them again. My thanks to the archivists at both the National Archives and Records Administration in College Park, Maryland, and the George Washington University Space Policy Institute Archives. Both staffs were extremely helpful and effective.

I also want to thank the staff at the journal *Quest: The History of Spaceflight Quarterly*. Excerpts of my argument on Eisenhower's domestic perspective and his role in the creation of ws-117L were published in their special issue on the fiftieth anniversary of satellite reconnaissance (volume 17, number 3) in 2010. They took a chance on an unknown historian from Canada, and it has been deeply appreciated.

Dr. Marc Milner, thank you for being a good friend and confidant over the years, listening when I needed to talk and giving me lots of encouragement. Dr. David Charters, thank you for taking on the role of thesis advisor and supervising me throughout my research.

Nicholas, Rob, Jeff, Andre, Dave, and the entire circle of friends who have journeyed with me on this project, you kept me sane and on an even keel, gave me useful advice, and kept me laughing. Thanks, guys.

Of course, my family: you deserve a great deal of credit for all of this. Supporting me when I was ready to give up, keeping me on course and working steadily, you have all earned a portion of the credit for this volume.

Finally, one other person stands out. Dwayne Day, you are a scholar and a gentleman. When my research looked about to founder because of a lack of material in the National Archives, you opened up a whole new set of doors for me, which included your own personal document collection. Considering we had only talked on the phone, that was an incredibly generous gesture. I can only pray that I will be able to repeat the favor for another grad student in the future.

INTRODUCTION

Filling in the Gap

When Americans went to bed on October 3, 1957, little did they realize that the night sky was about to change forever. The only forewarning was a small article in the *New York Times*, "Soviet Expert Tells West of Test of Rocket." Found at the bottom left corner of the first page on the morning of October 4, the piece would have garnered only passing interest had it not been for the events of the day.

At Tyuratam in Kazakhstan, just minutes before midnight on October 4, a Soviet rocket blasted off. At its top sat the Soviet Union's contribution to the satellite program of the International Geophysical Year (IGY). Weighing only eighty-four kilograms, the polished metal sphere of *Sputnik*, meaning "traveler," was an oversized radio beacon. Its sole purpose was to orbit the earth and make electronic noise. And what a noise it made! It permanently altered the tranquil night sky. The earth and the moon were no longer alone. A man-made signal pierced the eerie silence of space.

Americans did not immediately grasp what had happened. Traveling at over seventeen thousand miles an hour, the satellite (complete with its booster and protective shroud following in its wake) flew over the United States twice before the U.S. government became aware of it. Radio Moscow first broke the news to the world, playing up the achievement for its propaganda value. In the United States the shock and disbelief were evident. Believing strongly that the nation had to act and that the Americans' IGY counterpart to *Sputnik*—VANGUARD—would not provide results in the near future, Werner Von Braun, long a vigorous advocate of satellite research and director of the Development Operations Division of the Army Ballistic Missile Agency (ABMA), felt that all he needed was the opportunity to launch a satellite. He saw the Soviet success as a challenge that the United States had to meet immediately.

Most of the rest of the country reacted with shock and fear. The event shook Senate Majority Leader Lyndon B. Johnson (D-TX), who quickly realized its political implications. Following dinner on the evening of the 4th, he led guests at his Texas ranch outside to stare at the dark sky, which was no longer reassuring but now created a sense of dread. The consequences for U.S. national security were very clear. If the Union of Soviet Socialist Republics (USSR) could put a satellite in orbit, the same rocket booster could theoretically deliver a nuclear warhead to an American city. That prospect held enormous potential as a means of challenging Dwight D. Eisenhower's Republican administration (1953–61).

Media reaction was swift. Newspapers featured numerous articles that fueled further apprehension. While families gathered around their radios and televisions to listen to the insistent beeping of the new satellite; the print media sought to explain the event and its implications, while at the same time feeding public anxieties. If the Soviet satellite was heavier than the one the United States would launch, obviously the USSR was further ahead in space technology. By October 6 the press had linked recent cutbacks in U.S. defense expenditures to the USSR's victory in the first leg of what people would soon call "the space race," and the Eisenhower administration, with its goals of balancing budgets and restraining defense estimates, became the scapegoat. On October 8 Democratic senators entered the fray, demanding investigations and action. They accused the administration of complacency and of withholding funds from the satellite effort.

The Soviet's *Sputnik* advantage evaporated in August 1960, when the United States launched *Discoverer XIV*. Officially a scientific satellite, DISCOVERER was in fact the cover name for CORONA, the world's first functioning spy satellite. In less than three years from its beginning in early 1958, the United States had leaped well ahead of its chief rival in the military use of space for aerial reconnaissance. This gave it a significant edge—one that it maintained and expanded throughout the cold war.

This book seeks to explain how and why the United States surpassed *Sputnik* in only thirty-four months. It is remarkable that the United States was able to overcome all the technical challenges to orbit a reconnaissance satellite in such a short time. How was such a complex program able to succeed so rapidly? The answer rests with the work done on space programs from 1945 through 1958. The main focus is on the pre-CORONA program known as WS-117L, developed during the early years of the Eisenhower administration, starting in 1953 and ending in 1960–61. Most recent scholarship has concentrated on the triumphant CORONA satellites. While mentioning WS-117L, scholars usually do so in a cursory way (see the appendix). But without knowing fully about WS-117L, we cannot grasp the story of CORONA, and to truly understand CORONA's success it is essential to examine the satellite program as a whole. It is also necessary to situate it within the wider context of Eisenhower's presidency, especially his handling of national security policy.

The WS-117L program was the world's first attempt to develop a spy satellite and as such broke major theoretical and technical barriers. However, we know very little about it, and discussions in the literature are inaccurate and severely limited. Why is there so little insight into WS-117L? Part of the reason rests with the nature of the subject. Secrecy shrouds such programs and reduces the amount of material available to most scholars. In the information vacuum that this created around WS-117L, leaked details from military and other sources, combined with a healthy dose of speculation, led to several works that weave information and myth into a narrative account. Much of the information is contradictory, confusing, and inaccurate, often merging details from later programs with the initial WS-117L.

There is also a lack of material in other sources. The Eisenhower administration started and prosecuted the program. While there is a large body of writing on Eisenhower in the White House that is well developed and demonstrates a growing level of sophistication and research over time, the absence of information on WS-117L is glaring. The reason rests with Eisenhower himself. A firm believer in security and secrecy, he refused to release information on intelligence matters, even when it was to his advantage. Thus he left little material on his role in the program and the reasons for its acceptance.

The problem of figuring out WS-117L becomes more intense when we learn that the one organization that many people assume was most active in spying, the Central Intelligence Agency (CIA), had no real role in the original satellite program. WS-117L was a U.S. Air Force (USAF) program, and the CIA hardly participated at all. It was only in February 1958, in the wake of *Sputnik*, that this changed with Eisenhower's decision to separate part of the original WS-117L satellite into a separate program named CORONA. The CIA took on the job of developing CORONA with the USAF and only at that point began to play a major role in satellite reconnaissance. Not surprisingly CIA historians have looked mostly at CORONA and are virtually silent about WS-117L. Unfortunately understanding WS-117L is all the more difficult because CORONA was far more successful, producing the first operational spy satellite system. Thus it attracts a great deal of research, in the process obscuring the original program.

This combination of forces has effectively masked the WS-117L program. In particular the fixation on CORONA, as well as lack of information and slow declassification of documents, has created a great many myths. Since WS-117L is the origin point of virtually all military satellite efforts, and satellite reconnaissance proved pivotal in the cold war, an effort to reassess it is essential.

By examining satellite reconnaissance prior to CORONA, this book is unique. It makes a valuable contribution to our understanding not only of military space programs but also of the history of the cold war and of the Eisenhower administration. The examination of WS-117L that I provide here is the first comprehensive study of the program. Any understanding of CORONA is incomplete without such material. The relationship between WS-117L, the cold war, and the Eisenhower presidency is also a vital element in the story. The entire spy satellite episode helped shape U.S. cold war policy.

So as the troubled nation nervously eyes the skies, the American satellite story begins.

ABBREVIATIONS

ABMA: Army Ballistic Missile Agency
ADC: Air Defense Command
AEC: Atomic Energy Commission
AFB: Air Force Base
AFDAP: Air Force Development and Planning
ARDC: Air Research and Development Command
ARPA: Advanced Research Projects Agency
BUORL: Boston University Optical Research Laboratory
CGS: Coordinating Committee on General Sciences, Department of
    Defense Research and Development
CIA: Central Intelligence Agency
CIG: Central Intelligence Group
CSAGI: Comité Special Anneé Geophysique Internationale 1957–58
    (Special Committee for the International Geophysical Year)
DCI: Director of Central Intelligence
DoD: Department of Defense
DPO: Development Planning Objective
FY: fiscal year
GNP: gross national product
ICBM: intercontinental ballistic missile
IGY: International Geophysical Year (1957–58)
IRBM: intermediate range ballistic missile
JCS: Joint Chiefs of Staff
JPL: California Institute of Technology's Jet Propulsion Laboratory
JRDB: Joint Research and Development Board
kw: kilowatt

Mc: megacycles

NARA: National Archives and Records Administration

NATO: North Atlantic Treaty Organization

NIE: National Intelligence Estimate

NSA: National Security Agency

NSC: National Security Council

ODM-SAC: Office of Defense Management–Science Advisory Committee

R&D: research and development

RCA: Radio Corporation of America

RDB: Research and Development Board

SAC: Strategic Air Command

SIGINT: Signals Intelligence

SNIE: Special National Intelligence Estimate

TCP: Technological Capabilities Panel

USAAF: U.S. Army Air Forces

USAF: U.S. Air Force

USN: U.S. Navy

WADC: Wright Air Development Command

WDD: Western Development Division

WS: weapon system

PART 1

Eisenhower's Delicate Balance

[1]

Truman and Eisenhower on the Cold War (1945–55)

A distasteful but vital necessity.
—President Dwight D. Eisenhower

War . . . always starts with a Pearl Harbor kind of attack. In an
atomic war the first attack, no matter how well prepared
for it we may be, will really be a disaster.
—Louis Ridenour

Virtually every history of satellite reconnaissance justifies the creation of the program by citing the need for U.S. intelligence on
Soviet military capabilities. The argument focuses on the growing atomic threat from the USSR in the early 1950s combined with
problems in penetrating Soviet security as the primary drives for
the setting up of the WS-117L satellite program. This interpretation
is logical. There certainly were serious problems affecting the gathering of intelligence on the Soviet Union from 1945 to 1953, and by
the time Dwight Eisenhower took office in January 1953, the lack
of information was a growing concern. But was the monitoring
of Soviet military developments the sole reason for the WS-117L?

In this chapter I look at the contrasting views of presidents
Harry S. Truman (1945–53) and Dwight D. Eisenhower (1953–61)
on the threats and challenges that the cold war posed to the United
States beginning soon after the end of the Second World War.

Harry Truman and the Cold War (1945–53)

The United States required intelligence in the light of increasing tensions with the Soviet Union. During Truman's years in the White

House the international situation changed dramatically. Truman and his administration expected the wartime U.S.-Soviet relationship to last into the peace, and it startled them to find the peace so short-lived. Joseph Stalin, determined to ensure the safety of eastern and east-central Europe from Western influence, established communist governments there, thus creating a sphere of influence around his vast nation. When the Americans responded by providing assistance (initially financial, but later military) to help contain what they saw as communist ambitions, the result was a cold war that polarized the world and lasted decades. The deterioration of U.S.-Soviet relations and the start of the cold war did not occur overnight but rather emerged slowly through the second half of the 1940s and finally became an *idée fixe* in the American psyche during the first three years of the 1950s.

With mounting evidence of a change in Soviet attitudes toward the United States and a growing sense of hostility from the Soviet Union, the Truman administration became increasingly aware of the need for strong intelligence about the threats confronting the country. Pearl Harbor was the best argument for more national intelligence. The most shocking and destructive single experience in American history up to that time, this event traumatized every living adult American. The attack was possible because of a clear intelligence failure, the heart of which was the American inability to monitor Japanese military movements and intentions effectively. Most American information came from code-breaking, especially the top Japanese diplomatic code, PURPLE. However, absence of military data from such signals and the huge volume of traffic left intelligence experts unable to interpret indicators of a possible attack. This inadequacy and the absence of an effective system for coordinating and forwarding information to key U.S. commands climaxed in Pearl Harbor, which propelled the nation into war.[1]

The memory of that fateful day was still very fresh after war's end. Americans often recalled with perfect clarity what they were doing when they heard about the attack and the feelings it created, even years later. The trauma of the event and the suspicion that poor intelligence was probably to blame for it found reinforcement in the numerous investigations into what happened on that

4

day. Starting soon afterward with the Roberts Commission, the military and civilian arms of government conducted eight inquiries. The last, the formal Joint Congressional Committee Investigation (November 1945–July 1946), produced forty volumes of material, including much of the testimony from other investigations. Next to President Kennedy's assassination on November 22, 1963, and the attacks on the World Trade Center in New York on September 11, 2001, no other event in modern American history has had such an impact.[2] It was the fear of another Pearl Harbor that underscored the desire for intelligence and for satellite reconnaissance.[3]

In 1945 the United States was the sole possessor of atomic weapons, but that monopoly proved to be a paper tiger. Although the government was drafting plans to use such devices against the Soviet Union, it had very few bombs with which to do so. Immediately after the war it had no need to increase its atomic stockpile rapidly, as it did not see the Soviets as a threat or believe that they had any such devices. Possessing only conventional and chemical weapons, the Soviet Union could not project its power beyond Europe, let alone attack North America directly. Lacking any bases in countries near the United States and possessing only relatively short-range aircraft, it could not appreciably threaten its chief rival in the near future.

Conversely, because the United States did have nuclear weapons, it thought that dropping a few from long-range bombers would quickly knock out the Soviets if such an eventuality became necessary. Since the heavy bomber was the only feasible means of delivering the weapons, the U.S. Air Force understandably concentrated on preparations for strategic bombing, which became its primary task and evolved into a "bomber mentality" with respect to procurement, training, and intelligence assessment.[4] General Eisenhower noted that proclivity before he entered the White House in 1953, by which time the air force's fixation had become the norm. Authors such as Lawrence Aronsen argue that a major reason for this was the air force's solid belief that the Soviets wanted war and that the bomber was their only means of devastating the United States.[5]

One clear legacy of Pearl Harbor was the assumption that future wars would begin with a surprise attack. Having seen the advantage that surprise gave an attacker and knowing that the Soviets would eventually develop atomic weapons, Americans viewed the example of Pearl Harbor with great unease. The pairing of an unexpected attack and the power of nuclear weapons seemed a nightmarish combination that would paralyze the victim's economy and government. A corollary to this was the notion of a distinct U.S. disadvantage in the new atomic era. The postwar tendency in democratic societies to downplay military preparedness suddenly became a potential danger. The destructiveness of a nuclear strike meant that lack of preparation and neglect of military abilities would bring rapid defeat to a democracy.[6]

Immediately after the war strategic thinkers began to assess the impact of the bomb on warfare and national security. Bernard Brodie, an architect of nuclear deterrence theory and an articulate spokesman for the role of nuclear weapons in peacetime, quickly grasped the weapons' implications.[7] Arguing that their vast power would render any attack devastating, he concluded that the age of defense was over: some bombers would always make it through the defenses. Their destructiveness would prevent the victim's buildup of sizable military forces after the initial assault. Thus the United States had to be constantly ready to wage preemptive war. No weapon system, however "superior," could guarantee strategic superiority. To Brodie the best solution was deterrence. The key to safety was the retention of enough nuclear weapons to convince a potential aggressor of the likelihood of massive retaliation.[8]

An integral component of deterrence was the acquisition of accurate intelligence on the Soviet Union. However, until 1947 the United States had no centralized structure for doing so. The Office of Strategic Services under Gen. William J. Donovan had run wartime intelligence gathering and covert operations. It was unpopular within the administration, however, and Truman disbanded it quickly at war's end.[9]

Increasing tensions with the Soviet Union soon forced the president to rethink his decision and establish a permanent intelligence agency. The first steps in this direction took place on January 22,

1946, a little over one month before former British prime minister Winston Churchill added the phrase *Iron Curtain* to the Western world's vocabulary. Truman authorized creation of the Central Intelligence Group, or CIG.[10] The CIG was to correlate, evaluate, and disseminate all intelligence relating to national security, and the primary target was the Soviet Union. Although a major step forward, the CIG did not last long. Increasing demands for intelligence, rivalry among the military services, and limited resources prevented it from being totally effective. Moreover the new cold war necessitated an ever-stronger intelligence organization. Thus in 1947 the National Security Act gave the CIG a permanent statutory foundation as the Central Intelligence Agency (CIA).[11]

One of an intelligence community's primary tasks is preparing intelligence estimates. Under the CIG its Central Reports Staff had handled this task, assisting the director in producing these evaluations.[12] In 1947 the Office of Research and Evaluation took on this duty for the new CIA.[13] Accepting Truman's geostrategic vision of national security, the CIG (and the CIA) looked at a variety of factors that affected the United States. "National security" in the broadest sense now involved more than just weapons; it included economic forces, political and ideological threats, and control of resources and industrial infrastructure. Thus the concept became much broader in the Truman era.

The primary focus remained the Soviet Union, although the danger came not just from tanks and atomic weapons but from elements that experts had never really considered before in great detail.[14] Seeing the Soviet regime as hostile in every way, U.S. officials perceived possible threats not just in Soviet military actions but in national and regional instability and weakness in various parts of the world. The Soviets could exploit this situation through political, economic, and psychological means to undermine potential American strength.[15]

As officials were preparing the first estimates, lack of hard intelligence about the Soviet Union quickly became apparent. Initial estimates, such as the report of October 1946, "Soviet Capabilities for the Development and Production of Certain Types of Weapons and Equipment," indicated the problems of predicting the Soviet

Union's capabilities. Noting that "any report of this nature is at best educated guesswork," the authors pointed out that "an estimate of capabilities ten years hence obviously cannot be based on evidence, but only on a projection from known facts in the light of past experience and reasonable conjecture." As a foundation for their assessment, the writers relied on their current estimates of Soviet scientific and industrial abilities. They also compared American experience and estimates of capabilities (both current and predicted) with past Soviet capabilities. They also obtained information from former Soviet prisoners—mainly German scientists the Soviets had captured during the war and forced to work for them and who were now going home.[16]

Such estimates were problematic. First, by relying on American experience, they failed to account for the different setup of the Soviet economy: a "command economy" could call on more resources and use different avenues of research and development, thereby speeding development. Assuming that their own country's research and development was normative, American analysts found the Soviet Union to be far behind. Such a conclusion assumed that the Soviets would follow the same steps in the same order and in the same amount of time. This fallacy—"mirror imaging," as the historian Abram Shulsky describes it—involves "assessing or predicting a foreign government's actions by analogy with the actions that the analyst feels he (or his government) would take were he (or it) in a similar position."[17]

Second, this misinterpretation helped to create another problem: underestimation of Soviet capabilities, in which the absence of hard data played a role. Viewing the Soviets as backward, many Americans (including Truman) rejected the idea that the Soviets could compete in highly scientific and technical fields such as atomic energy. Noting the Soviet Union's postwar rebuilding and "limited technological development," the CIG's report argued that the USSR would be incapable of research and development in many advanced areas such as nuclear weapons for at least ten years.[18] Most predictions for the period up to 1950 attributed any Soviet achievements to captured German technology and scientists. The CIG anticipated that by 1948 the Soviets would only develop bombers with performance characteristics similar to the B-29s they had captured and interred in Asia during the war. It suggested that by 1950 they might produce almost 150 aircraft per month.[19]

Central to U.S. national security, of course, was the Soviets' ability to attack the United States directly with atomic weapons. While acknowledging the Soviets' overwhelming conventional strength, particularly in Europe, the CIA did not believe they were willing to risk open war in the face of the U.S. nuclear arsenal (at least in 1948).[20] The Soviet Union would require enough atomic weapons and an effective means of delivery, neither of which it possessed in 1948.

The accepted opinion of the U.S. scientific community in the period 1946–49 was that at worst the Soviet Union was five years from developing its own atomic weapon. Members of the Interim Committee who advised the president on postwar atomic energy concurred. Secretary of War Henry Stimson had set up the group in the spring of 1945, and it consisted of Vannevar Bush, James F. Byrnes, and James B. Conant.[21] The president ignored its prediction of a U.S. monopoly lasting only three to four years, as did many of his key advisors, who wanted to be the only player for a longer period of time.

Gen. Leslie Groves, director of the Manhattan Project, was by May 1945 sure that the United States and Britain had a monopoly on the crucial ingredient: high-grade uranium. Basing his reasoning on a special study of the world's uranium deposits called the Murray Hill Area project, conducted for him by top experts from 1943 to 1945, Groves believed that the Soviet Union lacked uranium and this would keep them at least twenty years behind in their development of nuclear weapons. No one ever challenged the highly secret findings, and Groves's committee had excluded any experts who might have dissented. As a result many U.S. officials thought that their country had gained control over the requisite raw resources. All of these experts and officials seriously underestimated Soviet capabilities.[22]

This type of complacency and overconfidence was common among U.S. officials, including the Joint Chiefs of Staff (JCS). Truman and his administration did not expect a Soviet nuclear device before 1950 or even 1955.[23] Even the CIA did not anticipate an atomic test prior to 1953. In a July 1948 memorandum for the president the CIA made it clear that it based its assessment on American, British, and Canadian experience. The agency found no reason to expect a Soviet weapon before the 1950s. Noting that it was "impossible to determine its exact status or to determine the date scheduled by the Soviets for the completion of their first atomic bomb," the memo added that the Soviets' supply of fissionable material would allow for only between twenty and fifty weapons by 1955, depending on the date of their first atomic test.[24]

The nuclear bomb would be useful only if the Soviets could deliver it to its target. In the CIA's report of September 28, 1948, "Threats to the Security of the United States," the agency stated that it did not believe the Soviets could attack the United States directly except via one-way suicide missions, which they could not launch at a scale sufficient to cripple the United States. The CIA predicted that the Soviets would not present a palpable threat from bombers and possibly by launching short-range missiles from submarines until 1955.[25]

The Truman administration felt safe behind American technological superiority—until 1949. The first Soviet test of an atomic bomb on August 29 of that year shocked the U.S. government, forcing a reversal in attitude about Soviet capabilities. This success antedated even the CIA's earliest prediction, and the threat of a nuclear attack would increase exponentially as Soviet weapon stockpiles grew. In fact by February 1950 the Joint Intelligence Committee (an interservice office that reported directly to the JCS) predicted that the Soviets would expand their atomic arsenal and would attack the United States "at any time they assessed that it was to their advantage to do so."[26] For the first time in American history, the country faced the prospect of a devastating attack that directly threatened its survival.

In April 1950 the CIA completed a comprehensive examination of the Soviet bomb's implications for U.S. security, and the results were not encouraging. It expected the Soviet Union to have approximately one hundred atomic weapons by 1953; the estimate climbed to two hundred for 1954–55. The Soviet version of captured American B-29 bombers (the TU-4, or BULL, bomber, as the West called it) was the expected delivery system. The report predicted that two hundred bombs reaching key targets could decisively cripple the United States.[27]

In June 1950 a further report, "The Effect of the Soviet Possession of Atomic Bombs on the Security of the United States," determined that the Soviet devices placed U.S. security in increasing jeopardy. Predicting a Soviet arsenal of between 70 and 135 warheads by mid-1953, the authors believed that the enemy could inflict critical damage. The greatest threat was "a single surprise attack on the United States and its foreign installations, which could seriously limit U.S. offensive capabilities, possibly to a critical degree."[28] More important, possession of the atomic bomb would greatly strengthen Soviet influence over other states by weakening their resolve to resist communism. Noting that the Soviet Union's basic objective was to establish communism worldwide, the report concluded that the Soviets would probably not start a general war unless they thought an attack by the West was imminent. However, if the balance of power began to shift in their favor, this attitude was likely to change.[29]

In November 1950 a CIA estimate, "Soviet Capabilities and Intentions," reinforced this interpretation. While adopting the same tone as earlier reports, it was more dire, predicting a probable general war sometime between 1950 and 1954 to install communist regimes in the West when the Soviets believed their strength to be at its peak. The bomb would be the major factor in the Soviet Union's estimation of its military power. Estimating that it already had 22 warheads, the CIA now predicted an arsenal of roughly 235 weapons by mid-1954. Equally important, the Soviets would have enough planes and trained personnel to deliver their entire inventory. American deterrence could no longer guarantee Soviet forbearance. The CIA expected that an attack would become more likely should the Soviets decide they could cripple or eliminate the U.S. arsenal in a single, decisive stroke.[30]

By early 1952 the agency had reversed its position. With the shock of the Korean War somewhat dissipating, it reported on January 8, 1952, that the Soviet Union was unlikely to initiate a general conflict even if it believed that it had the advantage, since its leaders preferred using any means short of war. The CIA did not expect the Soviets' development of intercontinental ballistic missiles (ICBMs) to change this situation. So long as they did not make a major technological advance, the threat of their attacking seemed very small; such a breakthrough could have led them to conclude that they could destroy the United States without sustaining effective retaliation. This meant not that nuclear war was impossible but that it was unlikely to be the product of a conscious choice.[31]

The shock of the first Soviet weapon was not the only one for the United States during this period. In September 1949 Chinese communists seized power and within weeks consolidated their control over continental China. Truman had hoped to contain communism within the Soviet Union and Eastern Europe by helping Western Europe and Japan rebuild, so the disappearance of the atomic monopoly and the "loss" of China shook many people in the administration and the public at large. In less than six months the United States went from "winning" the cold war to appearing to be on the edge of losing it. With ties between communist China and the Soviet Union strengthening, Truman and his administration were not ready for the next shock. Without warning, on June 24, 1950, North Korean troops poured across the 38th parallel into "democratic" South Korea. The Korean War drove home the problems facing the United States. Seeming to prove communism's lust to conquer by any means, the surprise invasion demonstrated that the U.S. government lacked solid intelligence concerning the intentions, plans, and capabilities of its greatest rival.[32]

The crux of the U.S. intelligence problem during Truman's administration was how to obtain the necessary information. The United States and the Soviet Union opposed each other politically and philosophically, and their societies worked differently. The United States was a comparatively open society, with freedom built into the Constitution and into the very fabric of everyday life. Freedom of the press, freedom of speech, freedom to travel within the United States, and relatively free access to information made the nation an open target for foreign intelligence agents. In October 1956 President Eisenhower observed that "a Russian can now buy an air ticket in New York and learn about our whole country" and suggested that there was little that anyone could do about it.[33]

In contrast the Soviet Union was a tightly controlled, "closed society"—in intelligence parlance, a "denied area." The Communist Party dominated every aspect of life. It restricted travel and thereby prevented its citizens (and anyone else) from seeing vast areas of the country. Using secret police and forcing people to spy on each other allowed the state to monitor its own people and foreigners. Harsh punishment and a climate of fear meant that few would provide information to U.S. agents or even make contact with them. These same controls restricted the activities of foreign visitors. Tight border security (complete with guards and fences) and the requirement of papers for domestic travel kept out most foreign agents or severely circumscribed their movements.[34]

The United States used every means possible to overcome these obstacles. Following war's end it gained intelligence mainly through indirect means. Captured reconnaissance photographs from the German Luftwaffe and the debriefing of German prisoners, émigrés, and defectors all provided some information, even if it was out of date and covered only a small part of the Soviet Union. Attachés at the U.S. embassy and occasionally tourists—both real and "special" ones (whom the CIA selected)—could provide more timely data, but the paranoid and heavy-handed regime monitored such individuals carefully to keep them from learning anything of real value.[35]

By 1948, in an effort to increase the amount of intelligence available, the United States under Truman's direction—and its allies—began to use camera-equipped aircraft for flights near and occasionally over Soviet territory. They also employed camera-equipped balloons as well as ground-, air-, and ship-based equipment to monitor electronic signals. The culmination of the balloon efforts was the 1956 program GENETRIX, "an Air Force meteorological survey" that used specially designed helium-filled balloons to photograph the Soviet Union in the first attempt to penetrate deep into its territory for intelligence purposes. Most balloons disappeared, and it was difficult to locate the subjects of recovered photos. U.S. analysts had never seen vast areas of the Soviet interior.[36]

Beginning in 1948 the United States also shared with Britain and Canada the task of intercepting Soviet signals. During the 1950s the National Security Agency (NSA) and its Signals Intelligence (SIGINT) and Communication Intelligence products remained the key source of information on the Soviet Union. However, because of excessive secrecy and difficulties in penetrating Soviet codes, SIGINT's value is difficult to assess.[37]

The reconnaissance efforts in this era inside the Eastern Bloc also began to include a human component. In the late 1940s the United States began infiltrating agents into Eastern Europe, equipping them with false papers, money, and transmitters and airdropping them into Soviet Bloc countries or rural Soviet border areas. A great deal of effort and money went into training these agents and preparing phony documents and histories, but few of these people provided any intelligence. The KGB and its network of informants apprehended most of them when they landed or shortly thereafter. Due to the large-scale failure of the program, all attempts to insert agents ended in 1954.[38]

The first Soviet atomic test in 1949 helped persuade the Truman administration to draft its most important document of the cold war: NSC-68. Finished prior to the Korean War but not formally accepted until after hostilities began, it called for increasing conventional and nuclear forces to counter the escalating Soviet threat. The Soviet test was also the key factor in American resolve to develop hydrogen weapons—a decision that escalated both the cold war and U.S. intelligence problems. The atomic bomb was a threat to U.S. security, albeit a relatively small one. Delivering it required long-range heavy bombers, whose slow speed would give a target nation hours to prepare for an attack, evacuate vulnerable populations, and perhaps intercept and destroy the delivery aircraft.[39]

In 1950 the ICBM as a means of delivering nuclear warheads had been a concept years ahead of its time, but its potential was already clear. Work had begun in the field immediately after the war; however, progress had been very slow. In his December 1945 report, "Towards New Horizons," Dr. Theodore von Karman, a noted mathematician and expert on aeronautical sciences and head of Caltech's Guggenheim Aeronautical Labs in Pasadena, predicted that it would soon become a vital new weapon system.[40] The atomic warheads' excessive weight, and the need for great accuracy to be effective, challenged researchers. The greater power of hydrogen weapons meant that smaller warheads were possible and the need for high accuracy became unnecessary.[41]

Eisenhower's New Approach (1953–55)

By the time of Eisenhower's election in November 1952, the government was acutely aware of the new reality of the cold war. It lacked reliable and accurate intelligence on the Soviet Union. None of its sources of information had alerted it to the pace of that country's atomic research in the late 1940s. Likewise, none of them knew about the state of Soviet research and development of its own hydrogen bomb, let alone warned of its test in 1953. U.S. intelligence was also unable to determine either the status of the Soviet bomber force (the most likely means of attack in the near future) or whether the Soviets had undertaken large-scale research in rocketry. This situation was coming to a head in January 1953, when Eisenhower became president.

The new chief executive took office worrying about an external military threat. Eisenhower's biographer Stephen Ambrose demonstrates that Ike suffered from the same shock regarding Pearl Harbor as his fellow citizens. It left a permanent mark on people's psyches, a mental sore spot, that made most U.S. leaders during the 1950s obsessive about the threat of another surprise attack. When atomic and hydrogen weapons made such an eventuality more feasible, many anticipated a new December 7 of epic proportions.[42]

Curtis Peebles supports Ambrose's interpretation. In *The Corona Project*, he argues that the new president faced two overwhelming problems that shaped U.S. policy throughout his years in office: overwhelming fear of a sudden Soviet attack and the inability of the U.S. intelligence establishment to penetrate Soviet secrecy, especially concerning military activity. He highlights Eisenhower's meeting with the President's Scientific Advisory Committee on March 27, 1954, when he explained his fear that modern (i.e., nuclear) weapons and a closed society gave the Soviet Union a major advantage.[43] Many historians take this reasoning as the basis for Ike's decision to support the creation of a system of satellite reconnaissance.

Eisenhower's military experience and knowledge made him well aware of the U.S. situation. Having experienced war and fearing nuclear weapons, he was probably the American most aware of the destructive potential of a surprise attack. Seeing war as "completely stupid and futile," he could find no way to decouple thermonuclear war from a conventional conflict. A "low-intensity" war would spread to include a general armed struggle between the superpowers and the use of nuclear weapons with nightmarish consequences.[44]

Briefings that the new chief executive received strengthened his fears. For example, on May 18, 1953, the Special Evaluation Subcommittee of the National Security Council (NSC) gave him its estimate of the scale of damage that the Soviets could inflict. Expecting attacks on bomber and forward-staging bases and on major population and industrial areas, the report painted a very bleak picture. The Strategic Air Command (SAC) could expect to lose between about one-quarter and one-third of its strength, and the country between one-third (in 1953) and two-thirds (by 1955) of its industrial output. Depending on timing, casualty rates ranged from 9 million people for 1953 to 12.5 million for 1955, half of them either from the immediate blast effects or from radiation exposure. The psychological impact would be unspeakable.[45] In the face of such estimates Eisenhower strongly opposed nuclear war except as a last resort.

For most of its history the U.S. heartland had been far distant from any potential enemy. In both world wars the Atlantic and Pacific oceans provided a great deal of protection. Now new weapons could eliminate the vast American industrial base rather quickly.[46]

Early in his first term Eisenhower spoke about the danger to Americans and the world in no uncertain terms. In his December

1953 "Atoms for Peace" speech to the United Nations, he pointed out that the United States could "inflict terrible losses upon an aggressor," but there was no real defense against nuclear holocaust. Eisenhower continued, "But let no one think that the expenditure of vast sums for weapons and systems of defense can guarantee absolute safety for the cities and citizens of any nation. The awful arithmetic of the atomic bomb does not permit of any such easy solution. Even against the most powerful defense, an aggressor in possession of the effective minimum number of atomic bombs for a surprise attack could probably place a sufficient number of his bombs on the chosen target to cause hideous damage." Eisenhower rejected the notion that the United States and the Soviet Union were "two atomic colossi . . . doomed malevolently to eye each other indefinitely across a trembling world."[47] By proposing a world stockpile of nuclear materials and internationalization of research on harnessing atomic energy for peaceful purposes, he hoped to defuse tension and preserve peace. Unfortunately this effort failed because of Soviet mistrust of American motives.

To Eisenhower, then, the only effective way to prevent nuclear war was increasingly vigorous intelligence gathering to provide warning of an attack and to reveal Soviet military capabilities. His wartime experience had taught him the value of accurate information.[48] When he took office, the scale of U.S. intelligence efforts surprised him. Three sources kept him constantly up to date: daily briefings on security developments, special studies and briefings by scientists and experts at his request, and formal National Intelligence Estimates (NIES) or Special National Intelligence Estimates (SNIES) from the CIA for the NSC.[49]

Overall the Office of National Estimates played a pivotal role in intelligence and in the president's and the NSC's deliberations. It provided about fifty NIES and SNIES per year, usually at the request of policymakers, synthesizing a vast amount of material into forecasts. NIES were only as good as the raw data that informed them. These reports directly influenced the U.S. defense posture and procurement as well as foreign policy.[50]

Growing military awareness of a crisis in U.S. intelligence was evident in an air force request of May 1953. That month Brig. Gen.

W. M. Burgess, deputy chief of staff for intelligence, USAF, admitted in an official study, "Cost and Effectiveness of the Defense of the United States against Air Attack in 1952–1957," that there was little information available concerning the Soviet order of battle. The problem lay in identifying the types and numbers of Soviet aircraft. The study reported that the turbo-prop Type-31 bomber was the only plane definitely in active service. It also predicted that Soviet versions of the American B-47 and B-52 jet bombers probably existed because the Soviets had near-complete access to American developments in military aviation. Since it assumed that the Soviet aircraft industry was simply copying foreign designs, the document concluded that the industry would follow the American lead and thus lag behind by only a few months. In short, the writers engaged in mirror-imaging.

By February 1954, however, Eisenhower began to doubt seriously that the Soviets could have any significant number of aircraft similar to the B-52, which was still under development in the United States. Basing his judgment on American experience, he believed firmly that the Soviets were not yet able to surmount the huge technical problems that the United States had already addressed.[51]

American concerns were also growing about Soviet research on long-range ballistic missiles. In 1953, lacking sufficient knowledge of Soviet work in this field, the Operations and Planning Group in the USAF Air Defense Command (ADC) turned to the intelligence community. Noting that the United States was preparing to meet a bomber threat, the ADC thought that such defenses aimed at aircraft-delivered bombs would be useless in the face of missile systems. Thus it desperately needed intelligence so that it could make appropriate preparations. It recommended maximum effort to obtain information and establish a factual basis for the evaluation of the status of and program for Soviet development of both long-range missiles and heavy bombers.[52]

Despite ongoing worries over deficiencies in intelligence, the CIA continued issuing reports without possessing solid information. In July 1953, for example, it released an SNIE on the expected Soviet capability of attacking North America. Focusing on the

period from mid-1953 to mid-1955, it did not assess the likelihood of such an eventuality but looked at how the Soviet Union
could attack *if* it chose to do so. Expecting that the Soviets would
continue to erode American superiority in atomic weapons, it
analyzed the principal means of delivery: long-range aircraft. It
estimated that as of July 1, 1953, one thousand TU-4s were available for attacks and noted sightings of heavy jet-bomber proto-
types in the air.[53] Projecting from American experience, the CIA
estimated two hundred heavy jet bombers available by 1955, sup-
plementing the projected two thousand TU-4s and about eighty
jet-powered medium bombers.[54]

The report is pivotal because it indicated that in 1953 there was
already evidence of the Soviet Union's developing long-range bombers. Also significant, it predicted that Soviet progress in long-range
aviation would resemble the American experience. It presumed
that Soviet leaders concentrated on heavy bombers and in-flight
refueling to permit maximum use of aircraft for North American attacks. But its most important feature was its speculative,
mirror-imaging character; its prognostications rested on virtually no hard data. It also discussed the possibility that the Soviet
Union was using the work of German scientists and captured missiles to develop long-range guided missiles.[55]

By early 1954, although intelligence efforts impressed Eisenhower, the limited amount of useful information that resulted dis-
appointed him. He found two major problems: reports made no
distinction between Soviet capabilities and intentions and seemed
lacking in perspective. The briefings and reports did not weigh the
Soviets' strengths and aspirations against the Americans' and did
not take into account American ability to counter Soviet bombers and retaliate in kind.[56]

These criticisms indicated serious problems. Failure to differ-
entiate between capability and intention was largely a function of
the data available. U.S. intelligence could monitor some technical
aspects of Soviet capabilities, and diplomats could photograph and
count aircraft during parades and aviation-day flyovers. However,
the United States lacked detailed knowledge about production
rates or capacities.[57] Extrapolating from this limited knowledge,

the United States had to estimate likely capabilities of Soviet long-range aircraft, their ranges, their bomb loads, and so forth. Nevertheless capabilities were the easiest factor to estimate.

The real problem lay in determining what the Soviet Union was planning to do. Was it going to build a large bomber fleet, and if so, why? Did it intend to create a bomber force so massive that it could overwhelm U.S. defenses, or was it simply mimicking the U.S. bomber mentality? Were the aircraft seen just prototype aircraft? The challenge faced by U.S. intelligence was how to answer these questions. The United States had no high-level spies inside the Soviet leadership, rendering it almost totally unaware of Soviet decision making and plans. It knew nothing about the Soviet mind-set. American assumptions rested on the interpretations of those people, such as U.S. diplomat George Kennan, who had had significant experience with Soviet leaders. Kennan's "long telegram" of 1946 had defined American views, even though he was an outsider within the USSR. Without access to the inner sanctums, U.S. intelligence could only guess at Soviet intentions.

Furthermore, without direct entrée to Soviet production facilities or airfields, and lacking accurate knowledge of production rates, plans, or decision-making processes, U.S. intelligence could only estimate vaguely the rate of aircraft production and deployment.[58] Military self-interest at home, particularly in the USAF, often skewed interpretation of findings in what one of Eisenhower's advisors called "sales promotion intelligence."[59] Playing up the need to enhance security, the air force used the scarce information available on Soviet bombers to maximize the perceived threat. It pioneered use of the "worst-case scenario" as the normative standard for intelligence reporting. Inflating the danger of course facilitated increases in deterrence requirements and thus budget requests, thus playing into competition for funding within the administration.[60] As late as 1956 Adm. Arthur W. Radford indicated to the president that debate over military programs reaffirmed the necessity for firmer intelligence estimates.[61]

The phenomenon of the "bomber gap" graphically illustrates this problem. In 1954 and 1955 the CIA and the air force were presenting increasingly bleak intelligence. Focusing on long-range bombers and citing studies by Albert Wohlstetter and other experts at the RAND Corporation concerning SAC's vulnerability, they predicted that the Soviet Union would launch a surprise attack when it had sufficient might.[62] Initial estimates offered modest projections of Soviet bomber strength. In April 1954 the TU-16 medium bomber ("Badger") and the new Type-37 four-engine heavy jet bomber ("Bison," first seen in July 1953 on the tarmac of the Ramenskoye test facility) appeared in Moscow flyovers, with between twelve and twenty TU-16s at a distance. The next month photographs of both TU-16s and Type-37s appeared, leading to speculation that the Soviets had roughly forty Bisons in service.[63]

The CIA saw the Bison as a prototype, with expected production to begin in 1956, and concluded that it would not pose a strategic threat until about 1960, but the air force did not agree. By June 1954 estimates of Soviet bomber strength had increased. In NIE 11-5-54 (June 7, 1954) the expansion (and corresponding ability to deliver nuclear weapons) seemed significant enough to change the world's balance of power. The almost simultaneous appearance of Badgers and Bisons was surprising, and so predictions of the Soviet Union's bomber strength escalated further. Predictions were now of 20 Badgers by mid-1954, 120 in mid-1955, and about 600 by 1959.[64]

In 1955, during practice flybys for the May Day parade in Moscow, observers spotted as many as ten Bisons in various formations. American experience suggested to the USAF that between twenty-five and forty could already be in service. Thus the USAF concluded that there was a gap between U.S. and Soviet bomber strengths and it was decidedly in the Soviets' favor. This quickly became a major political issue. Leaking these estimates to the media during testimony before Congress, air force leaders cited the Soviet bombers to help justify more military spending and thereby put a great deal of pressure on the administration.[65] The air force emphasized the worst-case scenario and predicted that the Soviets would maximize the rate of bomber production. U.S. experience suggested that the Soviets would have started with six aircraft per month, so at least thirty were already operational, if the estimate of the size of Soviet production plants was correct and if one assumed a maximum pace for production.

In hindsight there was indeed a gap, but it was in favor of the United States. During the second half of 1955 the United States began producing the first b-52s for training purposes, taking possession of roughly eighteen. By December 1956 it had 1,470 bombers for operations, including forty-five b-52s. The Soviet Union had only forty operational bombers in 1956, half of them Bisons.[66] The flybys of the latter were a ruse, using the same aircraft over and over to confuse U.S. intelligence. Unfortunately this did not become clear until after mid-1956, when American u-2 aircraft were able to fly over most key Soviet bomber fields.[67]

Although all the U.S. military services had access to the same information about Soviet long-range bombers, the air force supported the worst-case predictions: some eighty aircraft already operational and six hundred to seven hundred probably ready by mid-1959. In congressional testimony during 1956 and 1957 the air force maintained that by 1959 Soviet bombers would outnumber U.S. bombers by two to one. The army and the navy, along with the cia, were very skeptical of these figures. Only the air force saw the situation as absolutely dire; unfortunately its pessimism carried considerable weight with Eisenhower's critics and with supporters of a more powerful military.[68]

By October 1956 snies of Soviet bomber strength had again fallen victim to the air force's most unnerving prognostications. Presuming that Soviet goals included neutralization of U.S. retaliatory capacity, the air force whipped sketchy information into the prediction that the Soviets already had 1,400 bombers in mid-1956. The air force expected this number to grow by mid-1960 to 1,500 bombers, five hundred of them Bisons. The army disagreed strongly on the grounds that the air force lacked compelling evidence, but the intelligence community could not disprove the claims. The obvious solution, from the air force's perspective, was an increase in the main American deterrent: heavy bombers. By emphasizing the threat the air force was attempting to goad Eisenhower into lifting spending restrictions. In fact there was no "bomber gap." By mid-1956 u-2s flying over the Soviet Union revealed to the president that bomber estimates were grossly inaccurate.[69]

Even as the alleged bomber gap dominated U.S. security concerns, the development of the hydrogen bomb (H-bomb) began to render the bomber obsolete. The U.S. government was well aware that the value of atomic weapons was on the decline. In November 1952 a test at Elugelab in the Pacific produced the first hydrogen explosion, with a total yield of 10.5 megatons. The scale of destruction that this test implied startled experts. They had predicted that within fifty square miles of "ground zero" there would be death rates of probably 100 percent and within three hundred square miles massive destruction.

The nature of the cold war had changed yet again. RAND studies soon predicted that only fifty-five hydrogen warheads of a 20-megaton yield could totally destroy fifty of the largest Soviet cities and kill over 35 million people in minutes. The detonation of the largest American thermonuclear weapon on March 1, 1954, yielding 15 megatons, only reinforced the shift. The United States had the advantage in atomic warheads, but the H-bomb would limit the utility of its nuclear arsenal. Like HMS *Dreadnought* in 1906—an advance in military technology that made all capital-class warships preceding it obsolete—the H-bomb made atomic weapons useful but of lesser importance.

The H-bomb had startling repercussions for the ICBM program. In April 1946 the Convair Corporation had received a contract for $1.4 million from the U.S. Army Air Forces to study what was required for a ballistic missile with a range of up to five thousand miles. The government canceled the resulting MX-774 program in 1948 because of public opposition to defense spending and the seemingly insurmountable technical problems. Reactivated in 1951, the MX-774 program became the ATLAS missile system, but it was years away from becoming operational.[70]

The H-bomb changed this sluggish pace. The Strategic Missiles Evaluation Committee, also known as the Teapot Committee, first met in November 1953 to study the impact of the H-bomb on ICBM development. Working under Dr. John von Neumann, the committee acted quickly, submitting its final report in February 1954. The Teapot report clearly indicated that the new generation of hydrogen weapons exposed the United States to greater danger. The lighter warheads and greater destructive yield would make it easier to surmount the two greatest problems in ICBM development: thrust and guidance. A smaller rocket could launch a lighter warhead and was easier to manufacture and deploy. The greater destructive power lessened the need for accuracy. Instead of having to be within meters of a target for maximum effect, the warhead could detonate several kilometers away and still destroy it. More important, the ICBM was faster, and there were no effective countermeasures or adequate warning systems. The resulting prediction was that it was the ultimate delivery system. Logically the Teapot report pushed for immediate, strong acceleration of ICBM programs under one command.[71]

An apparently increasing Soviet bomber threat was creating apprehension, and the Soviet H-bomb only exacerbated fear of a nuclear Pearl Harbor. The air force and its supporters, including Senator Stuart Symington (D-MO), a staunch critic of the administration's defense efforts, advocated larger budgets and more bombers in response. Looking to make political gains from the threat of Soviet bombers, the senator, among others, used anxiety about a sudden attack to advance the Democratic agenda domestically. In the 1955 report of the air force subcommittee of the Senate Armed Services Committee and in his public statements, Symington blamed the president and his defense policies for the seeming decline in U.S. air power.[72]

By early 1954 Eisenhower had been viewing the far greater bomber threat and the new ICBMs as a frightening reversal of U.S. strategic fortunes. The CIA was the administration's principal intelligence tool, and it would warn of an impending attack and monitor the world situation. But it was now obvious that its intelligence picture was incomplete, with no indication that the situation would improve. During a special NSC meeting on March 31, 1953, the director of the CIA, Allen Dulles, had admitted to the president "shortcomings of a serious nature" in intelligence.[73] A year later NSC paper 5408, on continental defense, acknowledged continuing inadequacies: "In view of the implications of nuclear weapons in the hands of the Soviet Union, greater knowledge of Soviet capabilities and intentions is essential for military and non-military measures to reach maximum effectiveness."[74]

Lacking meaningful intelligence on Soviet military capabilities, knowing the effects of modern war, and anxious that an unexpected attack or a stupid mistake could lead to nuclear war, Eisenhower felt that he was at a major disadvantage when it came to selecting an appropriate level of defense preparedness.[75] Fearing that the situation could only worsen, the president turned for help to the scientists on the Science Advisory Committee, part of the Office of Defense Management. During a meeting with them on March 27, 1954, he raised his concerns about a surprise attack and the problem of penetrating a closed society like the Soviet Union.[76] He asked them to prepare a detailed study of the issues. James Killian, president of MIT, headed the study group. The Killian Commission and its report transformed U.S. intelligence operations, especially with its call for the more active application of science to intelligence gathering. The results of this effort included both the U-2 and satellite reconnaissance.[77]

As I mentioned, most historians see the need for intelligence as the driving force behind development of satellite reconnaissance. Historians such as Curtis Peebles and R. Cargill Hall cite the problem of gathering useful intelligence and the fear of surprise attack to explain the Eisenhower administration's move in that direction. The argument is not without merit. Both the president and the intelligence community wanted more information. The only antidote to growing anxiety about Soviet capabilities and intentions was concrete data from the CIA. The CIA's inability to provide timely and accurate intelligence necessitated a new means of gathering information. For this to happen two other elements were essential: political will and a viable alternative.

The arguments of Peebles, Day, Hall, and others focus, however, only on that aspect of Eisenhower's political agenda. By fixating solely on spy satellites' collection of intelligence, these historians have missed the wider implications of the decision to develop satellite reconnaissance. The will to act was the product of a variety of political forces, not solely the need for intelligence.[78]

[2]

Eisenhower and Defense

Three Challenges, Three Responses (1953–56)

> We face a hostile ideology—global in scope, atheistic in
> character, ruthless in purpose, and insidious in method.
> Unhappily the danger it poses promises to be of indefinite
> duration. To meet it successfully, there is called for, not so much
> the emotional and transitory sacrifices of crisis, but rather those
> which enable us to carry forward steadily, surely, and
> without complaint the burdens of a prolonged and
> complex struggle—with liberty the stake.
> —Dwight D. Eisenhower, "Farewell Address"

If he was to accept something as radical as satellite reconnaissance, Eisenhower needed convincing. The potential intelligence
from a spy satellite was a strong incentive for its development.
However, such a move involved a risky technological leap of faith.
After all, until *Sputnik* orbited in 1957, space flight was only theoretically possible, and few people thought of it as likely in the
near future—surely there were other, less challenging solutions
to the deficiencies in intelligence.

Why, then, would Eisenhower turn to space-based reconnaissance to address the issue? Satellites in fact promised the president something very valuable: a long-term source of high-quality
information on which to base U.S. policy. To understand Eisenhower's decision it is essential to grasp his thinking on defense
and national security. Throughout his two terms in the White
House, Eisenhower struggled doggedly with three major problems that plagued his administration, problems so systemic that
he believed they threatened the country's long-term safety. The

first two were products of the cold war itself and constituted his administration's primary policy challenges; he articulated these in an election speech in Pittsburgh on October 28, 1952. For him national security required balancing two massive tasks: defense of the nation's freedom against political and military disaster abroad and protection of the people against economic disaster at home.[1] Hence he saw his first task as protecting the United States from Soviet aggression. He accepted much of Truman's cold war philosophy, including the belief that the Soviet Union was aiming the communist world's formidable power and aggressive policy at undermining the United States.[2] He did not believe that the Soviets would resort to war but thought that the United States had to prepare for a cold war struggle of unpredictable length, probably decades.

His second problem—economic security—emerged out of the first. Facing a significant communist threat, Truman had dramatically increased defense spending; as a result rising taxes, inflation, and debt threatened the economy. Eisenhower realized that this could create far-reaching economic problems, undercutting the nation's well-being. He believed that a strong economy was the cornerstone of security policy and felt strongly that the country had to husband its economic strength, protecting itself by a level of defense spending that it could sustain indefinitely.

The third threat was the military's reaction to the first two. The military establishment and its supporters, both inside and outside of Congress, believed that the United States had to be ready to defeat communism at any time with overwhelming force. Lacking a clear intelligence picture of the Soviet Union, the military reacted by overemphasizing defense. Willing to accept any level of military expenditure, these pressure groups would fundamentally alter the country to save it. Eisenhower, however, believed that massive outlays on defense would help to create a garrison-state mentality that, in an effort to guarantee absolute security, would undermine the very fabric of American democracy.[3] The new president had to prevent overmilitarization without compromising national security. Any understanding of his decisions about satellite reconnaissance has to consider these three challenges—national security, economic security, and pressure from the military—and how he proposed to address them.

National Security: The Evolving Dilemma

At the start of his presidency Eisenhower faced the strategic situation that Truman left him. Like his immediate predecessor and most of his contemporaries, Eisenhower accepted many basic cold war assumptions: the Soviet regime was heavily armed, totalitarian, and hostile; it would try to expand its sphere of influence by any means possible; and its goal was world domination. He and Truman differed, however, over the immediate physical threat to the United States. Truman thought Soviet aggression would build toward an imminent attack during the mid-1950s. Eisenhower, in contrast, did not consider war inevitable; rather he saw the need to formulate a security policy that the nation could maintain over decades.

Truman formed his views on the Soviets during the tumultuous years immediately following the Second World War. Possessing no real background to help him evaluate the Soviet threat, but having experienced the shock of the news of Pearl Harbor, his administration struggled to adapt to the new cold war. In the initial euphoria over war's end in 1945, U.S. leaders hoped for peaceful coexistence; distance and the atomic monopoly helped them to feel secure. Unfortunately, when Truman was vice president (January–April 1945), President Franklin Roosevelt had kept him in the dark about U.S. agreements with the Soviets about postwar Europe. Expecting the Soviets to allow free elections in Eastern Europe, Truman felt shock at Stalin's insistence on maintaining a strong sphere of influence in that region, especially in Poland. Between the Eastern European issues and the apparent pressure by communist forces in Greece, Turkey, and the rubble of Western Europe, the U.S. president faced what looked like an escalating Soviet threat. Between 1946 and 1948 he attempted to devise a policy that would contain Soviet expansion and allow some measure of peaceful coexistence. NSC 20 (November 24, 1948) encapsulated U.S. policy, calling for a buildup in deterrent power while strengthening other nations to counter communist actions. The

1947 Marshall Plan to rebuild Europe and the Truman Doctrine pledging U.S. support for nations resisting oppression are excellent examples of the latter.[4]

As long as the United States had a monopoly on atomic weapons and the protection of distance, a direct Soviet attack seemed unlikely. But the Soviet atomic bomb of 1949, the "loss" of China to the communists that year, and the sudden start of the Korean War in June 1950 shook the U.S. government. Suddenly the Soviets seemed to be actively strengthening international communism. In response the administration reexamined national security policy and adopted a harder stance toward the Soviet Union. On September 30, 1950, Truman formally adopted the pivotal policy paper NSC 68, much darker and more dire than NSC 20: "The issues that face us are momentous, involving the fulfillment or destruction not only of this republic but of civilization itself."[5] According to the Truman interpretation that informed it, communism was inherently hostile to the United States and advocated a policy that was diametrically opposed to the American political worldview. Armed conflict between the superpowers seemed inevitable. The communist leaders, keen to expand their influence, would use any means available, violent or nonviolent, to subvert noncommunist governments. Truman saw his country as the only one that could arrest this influence.[6]

NSC 20 and NSC 68 differed on the scale of the threat. NSC 68 described Soviet nuclear capabilities as extensive. The authors, especially the principal writer, Paul Nitze, believed that the Soviet Union would attack the United States if it had the military advantage. As long as it believed that it could not attack successfully, the balance of power would preserve peace. Unfortunately Soviet progress was chipping away at the American lead in technology and research in such key areas as nuclear weapons.

Using available information, NSC 68 identified 1954 as the year when the balance would shift to the Soviets and an attack would occur. It insisted that the United States should seek to prevent this outcome by building up its strength, both military and economic, to provide a credible deterrent. It proposed that remaining more powerful than the Soviet Union would allow the Americans to hold onto the balance of power. Therefore it called for massive increases in defense spending and for notable corresponding growth in the deficit and in taxes to cover increased costs for national security.[7]

In the wake of the Korean attacks Truman embraced NSC 68. He authorized a rapid increase in spending in an effort to provide short-term security against the threat of Soviet attack by 1954–55. The annual defense budget went from less than $13 billion (approximately 5 percent of the gross national product, or GNP) in 1950 to $60 billion (18.5 percent of GNP) in the spring of 1951.[8] By 1952, according to Stephen Ambrose, Truman projected more than $50 billion for defense to prepare for the "year of maximum danger." But once the Korean War began to stabilize and a U.S.-Soviet war became almost a nonissue, Truman started to slow the military's expansion while keeping spending on it higher than prewar levels. The result was a series of budgetary deficits and a "feast-to-famine" pattern in defense spending.[9]

Eisenhower thought Truman's handling of the cold war flawed. The core of the problem lay in his predecessor's buildup of American strength to meet the perceived threat of a "year of maximum danger"—the exact opposite of what Eisenhower thought the appropriate response of a democracy. Instead of panicking, a democratic government prepares militarily only on a defensive (i.e., long-term) basis. As Eisenhower noted in his personal diary, "We do not attempt to build up to a D-day because, having no intentions of our own to attack, we must devise and follow a *system that we can carry as long as there appears to be a threat in the world capable of endangering our national safety.*"[10] Since the cold war was a long-term confrontation, no single day could hold greater risk for the United States; rather the threat would last, and a sustainable defense was essential for the long run.[11]

Eisenhower considered the notion of a "year of maximum danger" a fallacy in keeping with the traditional U.S. pattern of "boom-and-bust" military preparedness. He found that the idea collapsed in the face of logic and reason. The fixing of such a date was highly subjective and invariably shaped defense spending. When the threat seemed particularly serious, military expenditure grew, as in the wake of Korea and NSC 68, and when the threat appeared to recede or become less severe, it waned. Truman's approach had been disastrous, with the government ordering expensive equipment one year and then delaying or canceling orders the next. The dizzying oscillation of defense spending between 1945 and 1953 did more harm than good, preventing the military from modernizing and maintaining steady capability.[12]

While serving with the Joint Chiefs of Staff (JCS) in 1948 (and indeed as early as 1946), General Eisenhower had pushed for a constant level of spending—on the order of $15 billion per year—to maintain an adequate military. He anticipated that this amount would allow sustainable defense and steady development and modernization. At the same time, a fixed budget would permit (and force) the services to allocate their resources more effectively. In the long run such an arrangement would have helped to prevent inflation and to preserve the strength of the U.S. economy.[13]

During an April 1953 meeting with Republican congressional leaders, the chief executive, sick of criticisms that he would harm the air force by changing spending patterns, lashed out at those people who predicted that the country would be in more danger on any particular day, describing the idea as "pure rot" and adding that he had opposed this delusion all along. "I have always fought the idea of X units by Y date. I am not going to be stampeded by someone coming along with a damn trick formula of 'so much by this date.'" He argued that there was no way to achieve maximum military effectiveness in peacetime. That would require full mobilization of troops, which the United States could do only in war. There was no single perilous moment but rather "an age of danger."[14] The situation required an indefinite defense.[15]

Eisenhower's concern over the long-term nature of the cold war only grew with the presence of nuclear weapons in large numbers. His views on the matter were clear as early as 1946, when in a letter to his boyhood friend Everett "Swede" Hazlett, he described them as a "hellish contrivance."[16] They made war vastly more destructive, spreading its damage worldwide. For Eisenhower atomic weapons made general war obsolete because they had changed the nature of international conflict.[17] Conventional war was no longer feasible because, first and foremost, it would inevitably escalate into a nuclear holocaust. The advantages that came with the possession of nuclear weapons would create an overwhelming temptation to use them at the outset of a war in an effort to gain military advantage. The United States, however, did not have the luxury of employing them first. As a democratic nation it could not attack with such devices but could only keep them in readiness to counterattack and to paralyze the enemy at the start of hostilities.[18]

Following on the heels of this understanding came Eisenhower's second crucial insight into the nature of the cold war: for the first time ever war could devastate the United States. He tried repeatedly to convey the scale and implications of this destruction to his cabinet, the NSC, and members of Congress. In a short off-the-record speech in 1954 during a visit to Quantico, Virginia, he made it clear that there would be no victor in nuclear war. He could not bring himself to call the prospect of wiping out Soviet cities and capacity to wage war a victory. In the wreckage of that nation, where a government and functioning society had existed, there would be only a vast territory of devastation and ruin. The United States would have defeated the Soviet Union, but Eisenhower saw no political gains in creating such a wasteland. His views were clear: "Here would be a great area from the Elbe to Vladivostok and down through Southeast Asia torn up and destroyed without government, without its communications, just an area of starvation and disaster. I ask you what would the civilized world do about it? I repeat there is no victory in any war except through our imaginations, through our dedication and through our work to avoid it."[19] The devastating character of nuclear conflict had transformed the pace and tempo of warfare. This was strongly evident in the president's statements during a February 1955 meeting with leaders from Congress, where he dismissed the idea of sending troops to Europe after an attack. The vaporizing of American cities and infrastructure would rapidly degrade military capacity. The United States would not have the luxury of a protracted offensive buildup like that of 1941–44. The ability to marshal troops and to fight would disappear as moving, supplying, and replacing units (not to mention controlling them) became virtually impossible. Within the first few days both sides would tire themselves out. With destruction of cities from the air, it would be the task of the army to bring order to the chaos. The main role for the U.S. Army was first and foremost a domestic one: reestablishing control within the continental United States. Anyone who argued differently was "just talking through his hat. It couldn't be done and if I tried to do it, you would want to impeach me."[20]

By January 1956 Eisenhower was painting an even bleaker picture of the consequences of a nuclear attack. During an NSC discussion about stockpiling strategic materials in case of war, the president's tone left no one with any illusions. While attempting to describe the possible scale of destruction, Eisenhower lamented, "We were simply unequal to imagining the chaos and destruction which such a war would entail."[21] Rapid victory was out of the question: following the initial nuclear blows, he expected, the nation would be in ruins, with massive casualties and devastation of ports and cities. Although damage would be severe on both sides, the war would not be over. To ensure victory the United States would have to invade the Soviet Union and confirm that it could no longer fight, but repairing its own infrastructure and rebuilding its production to launch such an expedition would take at least three or four years.[22] In that time it would not be able to conduct conventional operations overseas.

In June 1954 Eisenhower set up an evaluation subcommittee to assess Soviet ability to inflict direct damage on the United States. This group—the chairman of the JCS, the director of the CIA, and several other key members of the administration—met periodically throughout his presidency. Its reports estimated Soviet damage under a variety of circumstances. The initial review provided both a worst- and a better-case scenario for a hypothetical attack on the United States at a predefined date. On January 23, 1956, some eleven days after the NSC meeting where the president had argued that it would take Americans years to dig out of the devastation, the subcommittee reported to him. A surprise attack on July 1, 1956, would, it projected, inflict shocking damage and destruction. Approximately 65 percent of the population would require medical care that would no longer be available. With total economic collapse and disintegration of the government, the country would be in a hopeless state. Even a one-month warning made for no sizable improvement. Although the United States expected to inflict three times as many casualties on the Soviet Union, the numbers were staggering to the president.[23]

In December 1956 the subcommittee again met with the NSC, and its projections for an attack in 1959 were even more dire. With both sides dramatically increasing their nuclear stockpiles, there would be massive damage. The United States could expect approximately 40 percent of its people to die and an additional 13 percent to sustain serious injuries and require medical attention at a time when it was scarce, if available at all. American retaliatory capability was the only factor preventing the Soviet Union from emerging as the dominant power within a day of the attack. By 1957 the forecast for an attack in 1960 showed no major changes, except for casualty figures: roughly half the people of both countries would die outright, with equivalent devastation to the structures of society.[24] As we saw, Eisenhower had a dim view of general war as a viable strategy. To Richard L. Simon of the Simon and Schuster publishing house the president clearly let his views be known: "War implies a contest; when you get to the point that contest is no longer involved and the outlook comes close to destruction of the enemy and suicide for ourselves—an outlook that neither side can ignore—then arguments as to the exact amount of available strength as compared to somebody else's are no longer the vital issues." Eisenhower pointed out that both sides would soon have sufficient nuclear strength to bring about total annihilation of one another: "Already we have come to the point where safety cannot be assumed by arms alone. But I repeat that their usefulness becomes concentrated more and more in their characteristics as deterrents than in instruments with which to obtain victory over opponents as in 1945."[25]

He saw the nuclear deterrent as a hopeful sign. Knowing that the United States could not initiate a war, he believed strongly that the Soviet Union would also refrain from doing so. From the start of his administration he argued that anyone—including Soviet leaders—who knew the weapons' destructive power would never use them. Although accident or miscalculation was still possible, the president thought it inconceivable that anyone would launch a nuclear strike. In July 1954 his administration took the position that increasing nuclear capabilities on both sides would make warfare less likely. Since total war meant total destruction, the very presence of nuclear weapons helped to prevent hostilities. By 1958 this stance had evolved: the nuclear deterrent rendered it impossible to imagine nuclear war arising from a conscious choice. If it did happen, it would be the result of an irrational act or accident, not of advanced planning.[26]

The nuclear deterrent had changed the nature of the cold war. While Eisenhower rejected Truman's view that a war with the Soviets was inevitable in the near future, he did accept that the United States needed to contain communism indefinitely. The indefinite nature of the cold war was the real danger. It required the United States to be ready to fight not on a particular day but perpetually if it was to maintain a credible deterrent. The cold war was a long, drawn-out struggle that could span generations, pitting the entire strength of both superpowers against each other. Eisenhower felt that he had to educate the military about this reality.[27]

The key to preserving peace was maintaining a balance of power. Rejecting the temptation to flaunt military strength, Eisenhower did not take a bellicose stand against the Soviets. He spelled out U.S. strategy in a memorandum of February 1953 to the NSC. The United States had to block Soviet expansion, decrease its power and influence, and develop the free world's strength to contain the Soviets politically and geographically. To do this the United States had to establish and sustain, for as long as necessary, a state of limited mobilization for war. This meant maintaining sufficient conventional military strength to deter Soviet aggression, developing the capability to mobilize rapidly in case of war, and backing it up with a nuclear arsenal second to none.[28]

In 1953 the administration conducted an in-depth study of policy alternatives to prepare itself better to wage the cold war. Eisenhower was a big supporter of study groups to provide detailed recommendations in response to specific problems. In May 1953 he formed Project SOLARIUM, consisting of a group of scientists and other experts, to examine three alternative courses of action and report back to the NSC. Each panel looked at a different approach to dealing with the Soviet Union. The first, Task Force A, was chaired by George F. Kennan and included key military and civilian figures, many of whom had helped to create the policy of containment.[29] Since 1948 this policy had called for the preservation of armed strength and a strong economy to deter communist aggression and expansion without increasing the risk of general war. Believing that the inherent contradictions and flaws in the Soviet Union would lead to its collapse, Task Force A focused on continuing containment. It called for a flexible policy to ensure that the United States could adapt to changing situations, but the danger was that the government would fixate exclusively on containing and destroying the Soviet Union. By emphasizing military containment of the Soviet Union, the United States ran the risk of losing its support from the free world, which feared another war. Therefore it had to focus not just on military preparations but also on psychological, economic, and political warfare.[30]

Task Force B included many figures who played a greater role in the Eisenhower administration, including Lt. Gen. James H. Doolittle, Douglas MacArthur Jr. (council for the State Department), and Maj. Gen. James McCormack; it examined a policy position that emphasized a more forceful stance on Soviet expansion. The United States should "complete the line now drawn in the NATO area and the Western Pacific so as to form a continuous line around the Soviet bloc beyond which the U.S. will not permit Soviet or satellite military forces to advance without general war."[31] Thus it had to maintain the military capacity to fight a general war against the Soviet Union for an almost indefinite period. Task Force B felt that this policy, though taking a hard line on Soviet expansion, meshed well with other policy positions and provided optimal flexibility.[32]

Task Force C included Lt. Col. Andrew J. Goodpaster (later a key aide in the administration), Lt. Gen. Lyman L. Lemnitzer (military planner, foreign affairs expert, and future supreme allied commander, Europe), and Frank G. Wisner (deputy director of plans for the CIA). Its report articulated the principle of "roll back."[33]

Accepting the other two task forces' approaches as a backdrop for a far more proactive policy, it advocated combining strengthening of the West to resist Soviet expansion (containment) with an aggressive political strategy that included covert, diplomatic, economic, military, and propaganda attacks on the Soviet Union. In the long term this policy would, it hoped, disrupt the Soviet Bloc's control over its territory, accelerate popular resistance, and exploit weaknesses within the Soviet Union—all in the cause of "rolling back" communism.[34]

Following the presentations of the three task forces in July 1953, Eisenhower summarized his understanding of these views to the entire group at Project SOLARIUM. He concluded from the project's deliberations that containment remained the best strategy for fighting the cold war. The only thing worse than losing a global nuclear war, he stressed again, was winning one; another war would destroy individual freedoms—a defining characteristic of American society. Thus U.S. policy must protect the nation and prevent open war. It is significant that the president emphasized the economic costs of defeating the Soviet Union. Looking at the long haul, he made it clear that demanding too much of the economy (and by extension the citizens) could lead to federal interventions in the economy that would erode individual liberty and rights. This was something that he was not willing to contemplate.[35]

SOLARIUM confirmed for Eisenhower that containing communism was the only practical, sustainable strategy.[36] He realized that the United States needed not overwhelming military force but rather a strong long-term deterrent, which nuclear weapons could supply. To be effective the strategy had to involve two elements: integration of atomic weapons into defense planning and development and a clear and straightforward policy for their use, about which the United States would inform the Soviet Union. How, then, to develop a strategy for use of suicidal weapons and yet make it credible enough for opponents to believe?

Eisenhower's solution was "massive retaliation." Secretary of State John Foster Dulles first spoke about it to the Council on Foreign Relations on January 12, 1954. A nuclear deterrent was to be the cornerstone of long-term U.S. security. The strategy played to American strengths. In Eisenhower's words, the United States "cannot be strong enough to go to every spot in the world, where our enemies may use force or the threat of force, and defend those nations."[37] Instead it would rely on a two-step solution. First, local forces, to which the Americans would extend limited aid, were to contain aggression as and where it occurred, and, when necessary, mobile U.S. reserves would join in the effort, along with naval and air support. By not committing massive ground forces, the United States would avoid frittering away scarce resources on hundreds of small battlefields. By building up the noncommunist world, Eisenhower expected to export some of the cost of a large military establishment and also expand the forces opposing communism.[38]

Second, the Americans would expand their own nuclear forces and continental defense. To prevent large-scale Soviet expansion and general war, they would turn their nuclear advantage into an umbrella of protection. The administration made it clear that when it faced such a crisis, it would react by using its greatest strength, long-range bombers carrying nuclear weapons against what it considered the "heart" of the problem: the Soviet Union. A preponderance of atomic weapons countered massive Soviet conventional forces. Knowledge of the devastating consequences of an attack in Europe, or elsewhere, would contain Soviet aggression.[39] For Eisenhower, then, the key to diminishing the risk of a large, general war was maintenance of a credible deterrent in the form of nuclear forces ready to attack the Soviet Union.

Economic Security: Stabilizing Defense Spending

The president's realization that the cold war could last indefinitely went hand in hand with his awareness that the military threat masked a more subtle danger to the core of American strength: the economy. He firmly believed that his was the best country in the world because of its individual freedoms and the material, intellectual, and spiritual opportunities that it provided for its people. Chief among the advantages it offered was the free-market economy. Eisenhower worried that the high cost of a military buildup might irreparably damage the U.S. economy. By expanding its defenses the nation might actually undercut its long-term ability to maintain them. This would force the federal government to make changes to the very fabric of society in the name of security.

The economy that Eisenhower inherited had experienced some dramatic upheavals since 1945. Truman had managed to avoid a severe postwar depression and had controlled unemployment thereafter, but the task had not been easy. Unemployment had increased dramatically, reaching 2.7 million by March 1946. As well, wholesale dismantling of the military reduced it by 1947 from over 12 million personnel in uniform to 1.5 million. These years also saw a series of clashes with labor. Demanding higher wages, automobile workers, steel workers, and coal miners went out on strike four times before Truman left office. Twice, in an effort to end a strike, the chief executive seized coal mines and the steel industry to restore production. Nonetheless he kept the economy expanding. Between 1946 and 1952 the GNP increased consistently, rising from $209 billion to $346 billion. A great deal of this growth resulted from deficit financing and the Korean War.[40]

For Eisenhower, who promised the nation "security with solvency," a major concern was the oscillation in defense spending under Truman. The defense budget for 1945 stood at $73 billion, or about 77 percent of the total federal budget. In the wake of the war and in response to strong pressure to demobilize, defense spending dropped rapidly. By 1947 expenditures hit an all-time postwar low of just over $9 billion, or about 24 percent of the overall budget. The figure for 1950 was slightly larger, at roughly $14 billion (33 percent of the federal budget). It was not until the Korean War and the adoption of NSC 68 on September 30, 1950, that the amount again began to soar. It jumped to over $33 billion in 1951 (74 percent of the budget) and reached $48.7 billion in 1953. And the defense budget for fiscal year 1955 that Eisenhower inherited was even larger: $46.3 billion, out of $78.6 billion. Truman left an overall national debt of over $275 billion. Thus since 1950 the greatest proportional increase in the national budget had occurred in defense, which by 1953 made up about 67 percent of the total budget.[41]

Eisenhower's views on the economy were fundamentally the

same as those of the rest of his party. Like most Republicans he subscribed to the concept of government and economic policy that Seymour Harris has described as traditionally Jeffersonian. Eisenhower believed in thrift, hard work, and minimal government interference in business. He felt that a government created the proper climate for prosperity when it kept taxes and spending low. According to the economic historian Craufurd Goodwin, Eisenhower, like all Republicans, avoided meddling in the economy too extensively because he believed that it was self-regulating. The more the government intervened in it, the greater the problems.[42]

The economy drew a great deal of Eisenhower's attention both before and while he was in the White House. Throughout the 1952 campaign, in a series of speeches, he hammered home its role as the foundation of American strength. In August 1952, addressing the American Legion Convention in New York City, he went to the very heart of the matter: "We must keep America economically strong. Even our great military effort must not break our great competitive system because in the combination of American spiritual, economic and military strength is the cornerstone of a free world."[43] One month later he went further, identifying the economy as a direct target of the nation's enemies. In October 1952 he again emphasized this point when he spoke in Jackson, Michigan, and in New Brunswick, New Jersey. Stating that "a free America must be the cornerstone of any free structure in the world," he argued that its people must maintain "our scientific strength, our productive and industrial strength, our financial strength to keep that economy sound. We must stay solvent."[44]

The danger was that the deficit of roughly $9.4 billion that Eisenhower inherited from Truman would lead to increased inflation. His predecessor had also authorized $81 billion in appropriations, and this increase in the nation's debt seemed a portent of economic stagnation. By July 1953 the administration had to borrow more funds to cover its predecessor's bills, raising the national debt to more than $272 billion and forcing the president to ask Congress to increase the debt limit.[45] Eisenhower had good reason to fear inflation: it decreased the dollar's buying power, thus making goods more costly. If the dollar bought less, the government would have to pay more for defense, which would increase the deficit or taxes. This in turn ran the risk of creating an inflationary spiral. Eisenhower believed that the government had to balance the budget so as not to increase the deficit and, where possible, to start paying off accumulated debt.[46]

The largest percentage of the budget went to national security, mainly the Department of Defense and the Atomic Energy Commission (AEC). In FY 1954, Defense, the AEC, and the Mutual Security Program used up about 70 percent of the budget, or roughly $50 billion.[47] Eisenhower worried that the country was living beyond its means, spending more on defense than the economy could support in an effort to gain "perfect" security. In his view controlling defense spending was the key element to balancing the budget.[48] The government had to find a workable balance between maintaining sufficient strength to counter possible Soviet aggression and ensuring economic solvency. Ideally this meant an austere defense budget in the range of $36 billion to $38 billion per year.[49] Eisenhower hoped to accomplish this sort of restraint with a combination of constantly upgraded, modern defenses, strong reserve forces available as necessary, and a vibrant economy. Failing to hold down defense spending would bring about a high risk of inflation and, eventually, severe government controls on the economy and a "garrison state" mentality. Since the armed forces existed to preserve a "way of life," not just property and territory, Eisenhower believed that to change the essence of the American system to defeat the Soviet Union would mean collapse in the long run.[50]

These ideas were not new to Eisenhower. From 1946 to 1948, while serving on the JCS, he had pushed for a constant level of spending. The Truman administration, however, under pressure to demobilize and reconvert the economy to consumer production, was slashing defense spending. In 1948 Eisenhower lamented in his diary that inflation had reached the point that even if the defense budget dropped to a constant $15 billion annually it would not be enough to meet U.S. security commitments. By increasing government spending through deficit financing and by raising taxes, the administration eroded the dollar's value and hurt the economy. As the dollar fell in value, the economy and the American people suffered.[51]

The danger that the cold war posed to the economy was a central feature of Eisenhower's discussions while onboard the uss *Helena* during his return from Korea in 1952.[52] He proposed a straightforward solution that balanced needs and resources, a middle course between rearmament and a sound economy.[53]

Economic considerations loomed large in nsc meetings, especially in Eisenhower's first year in the White House. His first State of the Union Address, on February 2, 1953, enunciated the issue succinctly: "Our problem is to achieve adequate military strength within the limits of endurable strain upon our economy. To amass military power without regard to our economic capacity would be to defend ourselves against one kind of disaster by inviting another."[54]

In a draft statement of February 1954 to the nsc about continental defense, he elevated the economic issue into an integral part of the struggle to win the cold war: "The survival of the free world depends upon the United States maintaining: (a) sufficient strength, military and non-military, to deter general war, to prevent or counter aggression, and to win a general war if it is forced upon us; and (b) a sound, strong economy, capable of supporting such strength over the long pull and of rapidly and effectively changing to full mobilization."[55] He also clearly laid out his position to senators during a private meeting in April 1956: "If we attempt to match soldier for soldier, weapon with weapon, etc., with the Russians, we will bring on ourselves at the very least economic suicide."[56] He insisted that "a bankrupt America is more the Soviet goal than an America conquered on the battlefield."[57] With defense portions of the U.S. budget so large, overspending had a major impact. The resulting rise in taxes diverted money from the productive private sector to the public purse, which was not productive.[58]

In response the president sought to correct problems in planning and spending for national security to preserve the economy. Foremost he dismissed nsc 68's conception of a "year of maximum danger." In a radio address on May 19, 1953, he spelled out the economic reasons for his rethinking of cold war strategy. He argued that the Soviets wanted the United States to overspend on security in order to precipitate an economic disaster. He called for reason and restraint in defense spending.[59] He thought that the defense program "must, first of all, be one which we can bear for a long—and indefinite—period of time." He would not accept a pattern of "sudden, blind response to a series of fire-alarm emergencies, summoning us to amass forces and material with a speed that is heedless of cost, order and efficiency."[60] Booms in spending on defense usually gave way to smaller budgets when the threat did not materialize; preoccupation with numbers was a hallmark of this sort of unrealistic thinking. Well aware that no fixed number of ships, planes, or guns (let alone dollars) could guarantee security, the president pushed for constant spending that guaranteed national security over an extended period without crippling the economy.[61]

Fluctuations in defense expenditures harmed the military budget and had far-ranging repercussions on the nation as a whole. The largest single component of the annual U.S. budget that the administration could adjust without legislative agreement was defense spending.[62] Its ups and downs affected taxes, the deficit, unemployment rates, inflation, and more. Thus preparing for war by a specific date was more disruptive than beneficial. Spending highs and lows affected the whole economy, and if war did not come by the appointed time, what then? Producing and training a large military force with very expensive equipment would have caused major interruptions to the economy. But once it was clear that war would not come, demobilization would follow, shifting production back to civilian uses and again disrupting the economy while it adjusted back to "normalcy."[63]

Eisenhower relied on his belief that the Soviet Union was not seeking a general war and pushed at every turn to make defense spending sustainable. Military expenditures, unlike consumer goods, massively drained valuable resources. In the domestic market, goods would spur economic growth; military items, however, were expensive and quickly became obsolete. Despite the income that military production generated, the goods that it produced were a drain on the economy.[64] In a speech before the American

Society of Newspaper Editors on April 16, 1953, Eisenhower elo-
quently linked the costs of large-scale defense buildups with their
domestic consequences:

> Every gun that is made, every warship launched, every rocket
> fired signifies, in the final sense, a theft from those who hunger
> and are not fed, those who are cold and not clothed. . . . The cost
> of one modern heavy bomber is this: a modern brick school in
> more than 30 cities. It is two electric power plants, each serving
> a town of 60,000 population. It is two fine, fully equipped hos-
> pitals. It is some 50 miles of concrete highway. We pay for a sin-
> gle fighter plane with a half million bushels of wheat. We pay for
> a single destroyer with new homes that could have housed more
> than 8,000 people.[65]

Anyone could understand this assessment of the cost of the arms
race. At best, large-scale expenditures on defense could help protect what a country had; at worst, they deprived people of basic
necessities.[66]

Eisenhower also disagreed with NSC 68's assumption that as
much as 20 percent of GNP should go to defense, supported by
deficit financing. He thought this ran contrary to common sense.
Instead he ordered his administration to start cutting expenditures and balancing the budget. Despite immediate and sharp cuts
in spending across the board, Eisenhower was realistic enough
to understand that achieving a balanced budget would take time.
The administration's first task was to slow the rate of spending of
the last Truman budget. This resulted in some confrontations in
April 1953 with Senator Robert Taft (R-OH) and other Republican
leaders over the failure to cut defense spending substantially. Taft
was particularly vocal about the need to reduce taxes and balance
the budget. Eisenhower pushed Defense to begin eliminating non-
defense costs, for example in administration, procurement, and
overhead; eliminating duplication and waste at this level would
save large amounts. The administration also began preparing a
comprehensive "inventory" of the entire defense establishment to
serve as a base line for understanding what the military had and
to act as a foundation for future cuts.[67]

The president realized, however, that cutting back on overhead and running Defense on an austere budget would not by itself stabilize defense spending. The administration turned to reorganization of the defense establishment in terms of the new realities of nuclear weapons and the urgency of economy. It started with cutting back on force size, keeping only key units, such as the elements of SAC necessary for retaliation and deterrence, at full strength. All units continued, but most of them with less than full strength—an unpopular move, particularly with the army. However, this was only the beginning.[68] On July 1, 1953, Eisenhower ordered the JCS to reexamine the whole defense posture with an eye to countering long-term Soviet aggression. This scrutiny covered not only strategic concepts but also roles and missions and many other areas, with emphasis on reducing costs. The resulting "New Look" was to maximize every defense dollar to ensure both security and affordability.[69]

Twice the JCS reported back to the president on this matter. On August 27, 1953, it presented its views and findings to the NSC. The four service chiefs had prepared this first look by the military at economy. Concentrating on budgetary issues, the report recommended reductions in forces abroad (especially in Japan, Korea, and central Europe) as cost-cutting measures. The United States would keep foreign bases but reduce the strength of overseas forces, transferring many of their tasks to local troops. The JCS believed that this could save money and ease the overextension in personnel; strategic and tactical nuclear weapons would offset decreased force sizes. The only real opposition to this document came from Gen. Matthew Ridgway, the army's chief of staff. He objected to the removal of forces from overseas bases and did not want the United States to rely solely on the nuclear deterrent and the air force. He felt that recent experience, as in the Korean War, suggested that it would be unwise to reduce the military's limited war capabilities and to rely solely on deterrence. Clearly he was also trying to protect the army's share of the budget.[70]

On October 13, 1953, there was a follow-up discussion of the defense budget for FY 1955. Since neither the Soviet threat nor U.S. security policy had changed and the administration had not issued a clear directive on use of nuclear weapons, Assistant Secretary of Defense Wilfred McNeil and the JCS presented a budget with few major spending cuts. McNeil predicted a defense budget for FY 1955 of $43 billion for combat forces alone; the budget for support elements was still being drafted. There would be probably about 3.5 million people in uniform. The administration's reaction was severe, with Secretary of the Treasury George Humphrey one of the most vocal critics. Predicting that the Department of Defense would spend $48 billion in FY 1954 and $47 billion (including carryover funds) in FY 1955, Humphrey insisted on cuts in the budget. The president reacted just as strongly, pointing out that the need for austerity meant that the military should be reducing its personnel ceilings and force sizes and reexamining its posture. For the military a key element in reducing spending was the extent of reliance on atomic weapons. Eisenhower was emphatic that the military should use these devices at any tactical level if an attack occurred. NSC 162, adopted a few weeks later, normalized their use if war started.[71]

Eisenhower insisted that Defense cut spending by about $4 billion in FY 1954 and by $6.6 billion in FY 1955. When the JCS refused to reduce expenditures, civilian leaders adjusted the budget before it went to Congress on May 7, 1953. The submission called for total spending of $43.2 billion for FY 1954, about $35 billion of it new spending. Overall this budget was below Truman's estimate by $5 billion. The air force lost roughly $4.7 billion, dropping to $13.7 billion, which forced a reduction in force size from 143 air wings to 120, only 114 of them likely to be operational by mid-1954. The navy's allotment fell by $1.7 billion to $9.8 billion. The army's actually went up, from $12.1 billion to $13.7 billion, to cover the final expenses of the Korean War.[72]

Such reductions occurred in virtually every Eisenhower budget. FY 1955 was to see new defense spending fall by $3.5 billion and previous expenditures by $4 billion. The overall defense budget was $37.6 billion (57.3 percent of the national budget). The army lost $4.6 billion and had to reduce its standing forces from twenty divisions to seventeen, for a loss of 317,200 personnel by mid-1955. Facing a decrease of 51,000 personnel owing to a loss of $1.5 billion, the navy cut its strength from 1,126 ships to 1,080. Reflecting Eisenhower's emphasis on massive retaliation, only the air force was able to expand. Rising from 955,000 personnel to 970,000, its wings increased from 115 in number in mid-1954 to 120 by mid-1955, with a projected goal of 137 by 1957. Brushing off protests from the army, the House Appropriations Committee cut the 1955 budget by an additional $1.1 billion, half of this from the army.[73]

This pattern remained through most of Eisenhower's two terms. The only exceptions were FYS 1956 and 1958, when Congress instigated small increases. The defense budget for FY 1961 stood at roughly $41 billion.[74]

The resulting New Look represented an attempt to readjust the U.S. strategic posture from preparation for war by a given date to sustained effort over the long haul. The U.S. advantage in nuclear weapons and strategic bombers became the focal point. By integrating atomic weapons into defense strategy down to the tactical level, and emphasizing their destructive power, the United States attempted to magnify their deterrent value across the full spectrum of operations, but it could now reduce its conventional forces. Although Eisenhower described the New Look as a reallocation of resources that emphasized nuclear deterrence, the budgetary savings were equally important.[75]

Unfortunately for the president, however, atomic and hydrogen weapons, although providing a means of reducing expenditures, were a double-edged sword. Soviet possession of the H-bomb, following their successful test on October 12, 1953, created anxiety within the U.S. defense community and the nation in general. The result was a greater desire for stronger continental defense. The support of the House and the Senate was important, but it was the military that felt the overriding need for greater efforts to prevent a "nuclear" Pearl Harbor. Eisenhower's belief that deterrence was the strongest defense, which sentiment the Defense Department echoed, did not find universal acceptance. More to the point, his initial attempts to reduce the defense budget put considerable pressure on the president.[76]

The New Look was a compromise between containment and financial overreach. Eisenhower predicted that the defense budget would level off by 1959 at about $30 billion—$35 billion per annum—less than Truman's projections but more than the department was spending before the war in Korea. The primary deterrent force consisted of Strategic Air Command's bomber forces, which could deliver large-scale nuclear strikes in accordance with the doctrine of massive retaliation. Conventional forces would become a mobile strategic reserve smaller than what most critics expected. At the same time, the United States encouraged Allied forces such as those in the North Atlantic Treaty Organization to be more active in their own defense, thus allowing greater U.S. reductions. To this end Eisenhower reinstituted spending caps on Defense to limit its share of the budget. In short, by relying on nuclear weapons he hoped to justify a more limited conventional force and save substantially in defense spending.[77]

Eisenhower solved the dual military and economic challenges of the cold war by creating a three-part, "layered" defense against Soviet aggression. First, the nuclear deterrent would contain the Soviet danger indefinitely. Second, the concept of sufficiency dictated U.S. deployment of enough forces (including atomic weapons) to augment local military forces in threatened areas. Nuclear weapons and local forces would provide adequate defense without damaging the U.S. economy. The third layer was the U.S. economy itself.

Pressure from the Military: Addressing Interservice Rivalry

Eisenhower's identification of the problems and the appropriate response did not result in an immediate solution. Rather they fueled a third major issue that haunted his presidency: chronic interservice rivalry over funding, which the cuts exacerbated. The president estimated in 1957 that he spent about two-thirds of his time fighting pressure from the competing services to increase spending on defense.[78] While he was looking at the "big picture" and the nation's long-term survival, the military focused on its primary mission: defense. The services' secondary, unstated mission—protection and expansion of their share of the budget—directly reinforced their principal task.[79]

Force reductions after war's end in 1945 had generated competition for limited funding. Each military branch played up its own ability to protect national security to ensure its continuing role and its need for more funding. Eisenhower was very familiar with the issue from his time as informal chair of the JCS. Seeing up close the rivalry between the army air forces and the navy over funding, he realized that Defense was causing many of its own problems. The larger its budget slice, the more important a service felt and the more prestige it assumed. The capacity to fulfill missions was not the issue; indeed the services were adept at creatively expanding their missions in an effort to justify more funding. Their tendency to run to Congress and the public with their convictions and complaints just raised the stakes.[80]

The problem became so severe by 1949 that General Eisenhower advocated drastic controls. First, the establishment of "majority rules" within the JCS would mean that all the services would have to come to a joint decision and then carry it out cooperatively. Second, the services had to realize that the president (i.e., Truman) was in charge. According to Eisenhower's personal diary, he felt that only the chief executive could resolve the issue: "I believe the President has to show the iron underneath the pretty gloves, some of our services are forgetting that they have a Commander in Chief. They must be reminded of this, in terms of direct, *unequivocal* language."[81] Eisenhower feared that the debates would one day erupt into a major confrontation that Washington papers and Congress would hear about, which would probably cause major problems in budget appropriations and in congressional committees. Only the president could prevent loss of control over the process.[82]

By 1949 the problems were too large for Eisenhower to correct. He felt that the members of the JCS so assiduously followed the doctrinal positions of their own services that they were unable to do their jobs, which involved a profound responsibility to the nation. The president and the secretary of defense would have to fix the problem.[83] The future president was also aware that the military was not averse to playing up perceived threats to increase funding.

In January 1952 his intimate knowledge of interservice rivalry led him to think carefully about the Truman budgets that called for $85 billion per year in defense spending. He saw the budget's size as a clear sign that the military was not doing its primary duty. Its obsession with spending and budgets and weapon systems had blinded it to other threats to the American way of life, such as economic stress.[84]

By the time Eisenhower entered the White House, the pattern of interservice rivalry was firmly in place. The new chief executive found within the military and other branches of government a tendency to approach everything from the perspective of bureaucratic self-interest. Looking narrowly at specific programs and special projects, or simply at increasing their share of the governmental pie, they were reluctant to cut favored projects (such as the atomic-powered bomber), scale back on spending, or think about the budget from the perspective of a team working for the nation. In fact the military was willing to use the very democratic process that it was defending against the administration. The services proved adept at public relations. They launched aggressive campaigns in support of particular projects or goals, manipulating the media through Pentagon leaks and press conferences.[85] Pet programs and even entire services thus appeared in the best possible light, occasionally at the expense of the other services, all in an effort to preserve or expand funding. Eisenhower wanted the public "to have a complete faith in the services—that is what he is working for." Unfortunately, because of the rivalry "the American public has lost a large measure of confidence in the services."[86] The competition had become so intense and overt that the public began to question whether the military could do its job.[87]

It was almost impossible for most people to judge the legitimacy of the armed forces' claims. Much of the general population had no real experience with the military and saw no reason to question its demands for new equipment or more forces. The military overcame objections of knowledgeable observers about procurement or budgetary issues by controlling the process of research and development. The first step in developing a new weapon system was demonstration of either a threat or a defined need. The source of the military's information was inevitably military intelligence, and that was a major part of the problem. Deputy Secretary of Defense Roger Keys argued in November 1953 that the services—especially the air force—were notorious for "sales promotion intelligence," or manipulation of intelligence to justify new projects. The practice resulted in large and overcharged programs of research, development, and procurement. The exaggeration implicit in worst-case scenarios accelerated requests for more equipment, personnel, and funding than was strictly necessary. The result was an overload of programs, which, once it started, was difficult to curb or cut.[88]

To help facilitate this the services maintained lobby groups in Washington, whose sole purpose was to press their case for the "needed" program. By 1953 Eisenhower had reports of up to seventy-five officials from the Department of Defense on Capitol Hill as military lobbyists, and he opposed their activities. The lobbyists reinforced the services' direct approaches to Congress. Going to staunch supporters such as Senator Stuart Symington, the air force tried to muster political clout to increase spending. By inflating the Soviet threat and overestimating the "weakening" of American defenses, the armed service lobbies applied considerable pressure for more spending in Congress. This activity disrupted Eisenhower's control over the military. The armed forces in essence was circumventing the normal chain of command.[89]

All four military services complained vocally about the president's budget cuts. By December 1954 Eisenhower was sure that there were too many military personnel who had no clear and useful purpose. He wanted to cut the noncombat people—the "dishwashers and waiters"—without sacrificing combat capability. Initially the most vigorous attacks on the administration's budget came from the army, led by General Ridgway. The emphasis on nuclear weapons over conventional forces gave the air force greater stature in national security, and thus the cuts fell heaviest on the army, and to a lesser extent on the navy. Army force levels dropped from 1.3 million personnel in December 1954 to approximately 1 million by June 1956.[90] As commander in chief Eisenhower had to judge what was necessary for the good of the country. It was up to his administration to establish what was essential for national security—a decision that it based both on the economy and on military strength. Ridgway and the top army brass strongly opposed the cuts and on several occasions raised the issue with the president. Emphasizing the damage to the morale of allies and to the armed forces themselves, Ridgway and Secretary of the Army Robert Stevens articulated what Eisenhower called the "parochial" view of the army; they argued for balanced U.S. military forces and less reliance on nuclear deterrence.[91]

The president's son, John Eisenhower, in 1972 gave an intriguing interpretation of the army's struggles for funding. He argued that his father was harder on the army than necessary to prevent charges of favoritism for his old service. The son was also critical of Admiral Radford (chairman, JCS) for taking advantage of his position to manipulate the situation and of Secretary of Defense Charles Wilson for his inability to control the military establishment. As for interservice rivalry, the younger Eisenhower pointed out that each branch handled relations with the White House differently. The air force, emphasizing its central role in nuclear deterrence, lobbied mostly in Congress, where many strong supporters from the struggle to create an independent air force still sat, and it had no hesitation about taking its position to the public. The navy used its contact with the White House to pass on its views directly to the president. Ironically the army had the least skill in public relations and tended to be more disciplined, so it was unable to fight cutbacks effectively.[92] Thus the air force was best at verbalizing criticism of the budget.

Not surprisingly the president was not happy with the air force. The central issue in contention was the number of aircraft in its arsenal. In 1953, taking its case to Congress and the public, the service claimed that it needed 141 air groups by 1954 in order to meet its responsibilities.[93] The president was furious. In conversations with Republican leaders he attacked these attempts to manipulate the budget: "I'm damn tired of Air Force sales programs. In 1946 they argued that if we can have seventy groups, we'll guarantee security for ever and ever and ever."[94] Now, under the guise of national security, the air force came up with a "trick figure of 141. They sell it. Then you have to abide by it or you're treasonous."[95]

Eisenhower was so angry partly because the air force was creating aircraft wings that were, in effect, paper tigers. Although it was building up to 141 wings, many of these units had aircraft but lacked trained personnel to fly or service them and facilities to maintain and operate them. The air force was overzealous in setting aircraft production rates and determining the lead time for procurement. Although the number of wings or the date when they would be operational did not change, the aircraft, associated equipment, and trained personnel were often far behind schedule. Eisenhower was willing to accept fewer wings as long as they were fully operational. His approach, in other words, did not involve a reduction of real strength; however, this did not satisfy air force supporters in the Senate.[96]

Senator Symington, the air force's spokesman on Capitol Hill, took up the fight for more money by attacking the administration. Charging that the cutbacks would leave the United States open to a Soviet strategic attack, he pushed for reinstatement of funds to the air force budget and for enlarging the air fleet to 141 air groups, an idea that Eisenhower totally rejected. As a general he had played the budget game, and he knew that the military overstated its needs, seeking thereby to build what Samuel Huntington called "political castles" to ensure its continued existence. Many people in the House and Senate had no background in such matters and tended to trust the military. Congressional leaders who knew better often accepted demands, especially when doing so was politically expedient or beneficial to their constituents.[97] Supporters of the air force in both houses tried unsuccessfully to increase its share of defense spending during debates on the budget for FY 1954.

Debates over the budget provoked even more heat in the years that followed. The budget for FY 1955 (prepared in 1954)—the first test of the New Look—generated fierce opposition from Symington and other air force supporters,[98] who eventually lost in a vote of 50–38. Later in 1955 General Ridgway (among others) aired his misgivings before Congress, arguing that cuts to the army jeopardized national security. Unfortunately for him the budget passed without major increases. Within days the air force released information concerning a possible Soviet increase in bomber strength. On May 17, 1955, Symington again took up the fight for defense spending. Suggesting that the nation might already have lost air superiority, he stressed that the Soviet threat was expanding in quality as well as in quantity. Pointing to Soviet bombers and hinting at the possible development of ICBMs, he attacked the administration's stance. Even when funds for bomber production increased slightly in the budget, that did not satisfy the senator. By June he was pushing again, this time to increase the number of air force personnel.[99]

Attacks by Symington and other air force proponents continued in 1956. Between April and July most of the testimony at hearings on air power before the Senate Armed Services Committee came from air force personnel asking for more funds. Gen. Curtis LeMay, commander of Strategic Air Command and a longtime hawk and advocate of strategic bombing, played up the Soviet threat. Predicting that by 1958–60 the United States would lag behind the Soviet Union in bomber forces by as much as 2:1, he warned that a surprise attack could eliminate the United States if the air force did not receive an additional $23.8 billion in the budget by FY 1958. A key forum for the "bomber gap" debate, these hearings (and those that followed the launch of *Sputnik* in 1957) clearly illustrate the problem facing the administration. During the budget hearings for FY 1957, LeMay used his professional knowledge of "additional unspecific information" about Soviet capabilities in an effort to add $3.8 billion to the air force budget. The danger that he presented was clear to the president: without some means of control, defense spending could rapidly increase without real justification.[100]

Constant pressure by the military and the tendency by many members of Congress to follow blindly posed a grave threat to national security, as Eisenhower saw it. Attempts to bring military leaders, especially the JCS, on board had failed. The chief executive was constantly dealing with officers willing to go behind his back to gain what they wanted. He went out of his way to convince the JCS and the military in general to accept the overall goals of the New Look. During a candid conversation in December 1954 with the joint chiefs and Secretary Wilson, he outlined the "big picture," stressing economic and fiscal policy. First, he based the defense budget on his study of security matters, and in his judgment it was sound. Second, as commander in chief he expected the loyal support of his subordinates for his decisions. If anyone had to complain, it should be to him in private, not to the media. Yet this appeal was to no avail.[101] As congressional attacks mounted over FYS 1955 and 1956, so too did Eisenhower's frustration. He expressed some of this to Swede Hazlett.[102] In 1955 he attacked the narrow military mind-set: "So what I need to make the Chiefs realize is that they are men of sufficient stature, training and intelligence to think of this balance—the balance between minimum requirements in the costly implements of war and the health of our economy."[103]

In an effort to curb Symington's calls for increased defense spending, Eisenhower took the unusual step of arranging three special briefings for the senator by Allen Dulles, head of the CIA. Symington left their first meeting, on July 21, 1958, skeptical that the Soviets were not ahead in ICBM development.[104] He based his opinion on information from Tom Lamphier, a former colleague and the assistant to the president of Convair, a major contractor for the missile program. Lamphier claimed to receive his insights from unnamed officials in the intelligence community who disagreed with the official CIA position. While most of the evaluation of Lamphier's intelligence remains classified, Dulles apparently believed that he obtained his information through his work at Convair.[105]

Dulles and the accepted opinion of intelligence circles could not put off the senator. On August 29 Symington wrote to the president detailing his assessment of the missile program, using information from Lamphier. He was emphatic that American efforts were inadequate. The intelligence community had underrated both the scale and the capabilities of Soviet missile development. The CIA had predicted roughly five hundred Soviet ICBMs operational by 1960–61; the United States would have only half that number by 1962. Symington questioned the CIA figures on Soviet missile tests, citing U.S. intelligence to show that American efforts were insufficient. He was confident that the United States was not doing enough for defense, and it obviously disturbed him that no one would agree with him.[106]

Again confiding in Hazlett the president outlined the threat that the relationship between Congress and the military posed. His personal knowledge and experience allowed him to counter many of their efforts, but he worried, "Someday there is going to be a man sitting in my present chair who has not been raised in the military services and who will have little understanding of where slashes in their estimates can be made with little or no danger. . . . If that should happen while we still have the tension that now exists in the world, I shudder to think what could happen to this country."[107]

Three Challenges, Three Responses

Historians have tended to explain the advent of satellite reconnaissance in terms of fear of a surprise Soviet attack and the military's need for information about Soviet strategic capabilities. Eisenhower was fully aware that such an event was a possibility that the United States had to prevent. However, this was not the driving force for the creation of the first spy satellites within his administration. Other concerns helped to motivate him to experiment with a radical new system for collecting intelligence.

The three-pronged threat—from the misunderstanding of the true nature of the cold war, the economic threat facing the nation, and pressure from the military—served as a major impetus for Eisenhower's acceptance of space-based reconnaissance. These were the concerns that were present from the beginning of his administration and dominated almost every decision and policy related to national security that he made while he was in the White House.

Eisenhower responded to these problems by adopting a nuclear deterrent, trying to stabilize defense spending, and addressing interservice rivalry. Massive retaliation, expansion of the nuclear arsenal, and incorporation of atomic weapons down to the tactical level provided (in theory) a strong deterrent to Soviet aggression. Soviet fear of a nuclear response was the big stick that the president expected to preserve American safety throughout the cold war.[108] Nuclear weapons, ironically enough, allowed the United States to wage the cold war indefinitely by providing a "cheap" counter to conventional forces. For them to be effective, however, the United States had to keep abreast of Soviet military capabilities to ensure that they maintained sufficient defenses.

The difficulties that Eisenhower encountered in curbing military demands demonstrated the importance of accurate intelligence. The president needed an extremely reliable means of gathering intelligence that would provide the information necessary to keep the American deterrent effective for several generations. Such a system was essential to determine sufficient force levels and to assess future military and economic needs during the cold war. Equally important, it could provide future presidents with information to counter military demands for ever-escalating defense spending.[109] A satellite that could constantly monitor the Soviet Union would be the ideal solution.

[3]

Eisenhower and Satellite Reconnaissance

Three Projects (1954–58)

Intelligence applications warrant an immediate program leading
to very small artificial satellites in orbits around the earth.
—Report of the Technological Capabilities Panel, February 1955

The need for information on the Soviet Union to manage the
cold war is one thing, but this need does not necessarily lead to
the development of spy satellites. In 1954 the idea of space-based
reconnaissance was at best something from a science fiction novel.
For it to become reality, politically the Eisenhower administration had to first be aware of the idea of spy satellites and it had
to take political steps to make it happen. Between 1954 and 1958
three major developments proved seminal to fostering and protecting overhead reconnaissance.

First, in July 1954, at the president's request, Eisenhower's Scientific Advisory Committee set up the Technical Capabilities Panel
(TCP), also known as the Killian Commission (July 1954–February
1955). In November 1954 it reported that the state of U.S. intelligence was distressing, and on its advice the president made a
visionary commitment to pursue the U-2 high-altitude aircraft
and space reconnaissance so as to obtain better information about
the Soviet Union.

The resulting TCP report directly shaped U.S. intelligence and
space policy from that point forward, and its importance cannot
be overstated. In terms of cold war policy, the report (sometimes
referred to as the Killian Report) has been described as the most
influential study to come out of the Eisenhower administration.[1]

Former CIA official Richard Bissell argued that this report played a major role in initiating the development and deployment of a series of reconnaissance systems that drastically expanded the scope of the whole U.S. intelligence collection process.[2] More notably the TCP report led to the merging of American space, defense, and intelligence policy. As a result the United States began to pursue specific objectives meant to guarantee continued access to intelligence gathered from space for decades to come. This intelligence became a major counterweight against Eisenhower's worst fears for the future of the nation and ultimately changed the course of the cold war.

Second, and following up on the TCP report, the United States began planning to provide legal protection for both the U-2 and spy satellites. In 1954–55 Eisenhower attempted to finesse the legal issues of overflight for intelligence gathering. To protect the U-2 program he enunciated his Open Skies proposal at an international meeting in Geneva in July 1955. To help establish the legal precedent that delineated the line where airspace ended and outer space began, a major issue in the final TCP report of February 1955, the United States announced its support for a small scientific satellite as part of the International Geophysical Year (IGY, 1957–58). He hoped to use the project to help legitimize satellite reconnaissance. This was formalized in NSC paper 5520 (May 1955), which enunciated a national satellite policy and committed the United States to launching an IGY satellite. This policy was augmented in the wake of *Sputnik* in June 1958 in the form of NSC 5814. A more comprehensive policy, it helped shape U.S. space activities for the next decade.

Third, the United States began Project VANGUARD, a scientific satellite in support of the IGY. Started in July 1955 as a completely civilian program, it was meant to establish the legal precedent in a way that would not antagonize the Soviet Union. Although it failed to beat *Sputnik* into space, it displayed the commitment of the president to solving the legal issues involved in space-based reconnaissance.

Together these efforts—the Technical Capabilities Panel, the IGY satellite, and VANGUARD—and the formal U.S. space policy chart a cohesive, long-term endeavor to enhance and safeguard U.S. intelligence gathering. They show the extent of the president's gamble in applying innovative methods to solve the problems that he faced. Yet in 1958 the very satellite that he was trying to safeguard was still in the drafting stage.

The Technical Capabilities Panel (July 1954–February 1955)

As president Eisenhower had ultimate responsibility for national security at a time when the nation lacked strategic intelligence on its major rival: the Soviet Union. In light of apparent increases in Soviet long-range bombing capabilities, there was cause for concern, and the matter aggravated the president's problems in managing national security. Already facing calls by the armed forces and some members of Congress to increase defense spending, particularly on the air force, Eisenhower realized that without adequate information he could not justify the expenditures or counter demands to increase them. Although he doubted that the Soviet Union would choose to attack the United States directly, should a mistake occur the lack of information could hasten a global disaster.

To clarify the issue and provide a long-term answer that would help guarantee security for generations to come, President Eisenhower turned in early 1954 to a group of noted scientists who sat on the Science Advisory Committee under the Office of Defense Management(ODM-SAC). The ODM-SAC was already well aware that the administration worried about a surprise attack. As early as March 15, 1953, it had warned of the country's vulnerability to such a strike. In the following year Trevor Gardner, assistant to the secretary of the air force for research and development, encouraged it to look more actively at defense policy; he also invoked his knowledge of RAND studies that warned of how vulnerable U.S. retaliatory forces were to maximum effect.[3] He pushed the Science Advisory Committee to play a more active role and alerted them to the president's concerns about American vulnerability prior to their meeting with him on March 27, 1954.[4]

When ODM-SAC scientists met with the president he revealed his concerns about gathering intelligence and the risk of surprise attack. Facing conflicting perspectives from the military and the CIA, he sought to outmaneuver them and asked the group to draft a "root and branch," detailed study of how to lessen that risk. Chair Lee DuBridge asked James Killian, president of MIT, to convene a subcommittee to examine the feasibility of such a study and to make recommendations.[5]

The Killian Commission (July 1954–February 1955)

Killian's subcommittee quickly convened to consider the matter and in less than a month reported back to the ODM-SAC. It recommended creation of a technical task force to study methods of countering surprise attacks, look at the technology underlying American weapons and intelligence, and pass on its findings to the White House. DuBridge recommended to Dr. Arthur Flemming (director of the Office of Defense Management) that he recruit such a group and that the president give it official authority and support. Three panels would examine three critical subjects: continental defense, striking power, and intelligence. The president approved the proposal on July 26, 1954, in effect giving these civilian scientists complete access to national military and intelligence secrets. Many military experts did not like the idea of allowing scientists to look at such sensitive information. The decision was particularly courageous in light of the paranoia about communist sympathizers and spies that emerged from the congressional hearings under Senator Joseph McCarthy (R-WI).[6]

DuBridge recommended Killian to chair the steering committee of what became the Killian Commission, with Dr. James Fisk to act as associate director. Knowing Killian from his own years as president of Columbia University (1948–52), Eisenhower readily accepted him as chair. Killian embraced the job wholeheartedly and quickly formed a steering committee, assigning personnel to various tasks. Some forty-one scientists, engineers, and military and communications experts joined sixteen military and civilian members of the government. The steering committee accepted the three-panel approach, to which it added several supporting studies in communications and technical personnel to round out the report.[7]

Overall the TCP report addressed five key goals. The first and foremost was development of American capacity to gain more intelligence about a potential enemy's intentions and capabilities—principally to receive warning of any planned attack. The second was increasing retaliatory power through the use of new technology, which might strengthen the deterrent force and decrease the risk of an attack. The third was to strengthen defensive capabilities in order to blunt any form of assault. Fourth was assuring secure and reliable communications. Finally came the need to understand the effect of technology on the military's personnel requirements.[8]

Land's Intelligence Panel

All three panels—on continental defense, striking power, and intelligence—proved vitally important, but the third had the most far-reaching impact on the cold war. It was not only instrumental in the decision to develop satellite reconnaissance but it was the major factor in spurring development of the U-2 aircraft. It also stimulated formulation of major policies relating to space and to the Open Skies proposal of 1955. It is this panel and its findings that I now examine in greater detail.

Killian chose Edwin Land to head the intelligence panel. A genius and an innovator, he was an unorthodox thinker and a solver of impossible problems. Creator of the Polaroid Corporation, with hundreds of patents to his name, he also had significant influence in Washington's intelligence circles. As a major player in the photography industry, he was a natural choice to advise the administration on aerial reconnaissance. In the words of his biographer, his approach to innovation was consistent: "Defining a need and the shortest path to a practical answer, he did not think there was a law of nature forbidding what you wanted or specified."[9] Land's six-member panel was relatively small for the task, reflecting his own style of research and his belief that a committee must be small enough to "fit into a taxicab" in order to avoid conflicts of will and personality. His panel was a veritable who's who of experts. It included James Baker (associate director) from the Harvard Optical Laboratory, the leading lens designer for aerial photography; Joseph W. Kennedy, a renowned chemist from Washington University in St Louis; Allan Latham Jr., a long-time friend of Land's and on the staff at Polaroid; Edward Purcell from Harvard, a Nobel laureate in nuclear physics; and John W. Tukey of Princeton and Bell Laboratories. This group, responding to Land's demand that the United States have the best intelligence system possible, received frequent briefings from various intelligence officials.[10]

The intelligence panel, in fact the entire Killian Commission found the state of U.S. intelligence appalling despite its best attempts to locate people in the CIA and the military who supposedly knew about gathering and processing such information. Years later, during an interview, Land recalled, "We would go in and interview generals and admirals in charge of intelligence and come away worried. Here we were, five or six young men, asking questions that these high-ranking officers couldn't answer."[11] The only individual who impressed the panel was the former naval officer Arthur Lundahl, who came over to the CIA in 1952–53 to head their photographic-interpretation group, which at the time relied on air force photos from both periphery and limited penetration flights of the Soviet Union. He was a strong and vocal advocate for gathering intelligence via photographing the Soviet Union.[12]

The commission's final report, which it submitted on February 14, 1955, sharply criticized the quality of intelligence. Echoing the president's concern about the matter, it noted, "We must face up to the reality of the actual situation: our estimates of the capabilities of the Soviets have, at their center, only a very limited core of hard facts, and we know even less about their intentions."[13] In one of the report's most forceful statements, Land wrote, "We must find ways to increase the number of hard facts upon which our intelligence estimates are based, to provide a better strategic warning, to minimize surprise in the kind of attack, and to reduce the danger of gross overestimation or gross underestimation of the threat. To this end, we recommend adoption of a vigorous program for the extensive use, in many intelligence procedures, of the most advanced knowledge in science and technology."[14] Land considered intelligence absolutely essential, thinking about it and its role in the policy process "in its most constructive and benign sense, as a search for the knowledge to reach sound national policies." Like the president, he believed it was vital for limiting or eliminating surprise attacks and for reducing the tendency to overestimate or underestimate a possible threat. Irrefutable facts would prevent the military or civilians in government from overreacting and allow instead for cool and rational decision making. Thus solid intelligence would allow the United States to "better cope with those fantasies about our weaknesses and the enemy's superiority that occur occasionally among the military or the politicians."[15] Clearly Land was thinking along the same lines as Eisenhower.

The TCP report stressed that the U.S. government should be the best informed in the world. This would not only protect the nation but also help to resolve internal differences concerning national security. More significant, "if intelligence can uncover a new military threat, we may take steps to meet it. If intelligence can reveal an opponent's specific weakness, we may prepare to exploit it. With good intelligence we can avoid wasting our resources by arming for the wrong danger at the wrong time. Beyond this, in the broadest sense, intelligence underlies our estimate of the enemy and thus helps to guide our political strategy."[16] The president liked this argument. His concerns about curbing the military's spending and his criticism of its demands and lobbying found support in the TCP report.

Recommendations of the Land Panel

Land and his intelligence panel made two sets of recommendations about increasing the quantity and quality of hard data for security. The first set appeared in the report's section on intelligence. These were official and "open" recommendations, emerging from the intelligence panel's insight and experience. The second did not form part of the report. The panel deemed this material too sensitive for the full National Security Council and discussed it only with the president in November 1954; they handed the president a single copy of these findings for his eyes only in February 1955. Together these two sets changed the direction of U.S. intelligence collection.

The Land panel's official, open recommendations were more

general and aimed at improving intelligence gathering without specifically defining methods or equipment. The most important for our purposes involved calls for increased use of science to penetrate Soviet security and prevent hoaxes that could mislead the United States. One urged development of safeguards to ensure that National Intelligence Estimates did not accidentally release their sources of information.[17]

Land's intelligence panel was especially worried about electronic intelligence gathering. Members of the panel felt that technical and administrative problems were severely hampering that effort. They wanted a speedy solution via "a combination of technical knowledge and adequate authority at a high level."[18] At a later stage both of these concerns had long-term repercussions for satellite reconnaissance. Within the administration stringent security and compartmentalized information on a "need-to-know" basis emerged in response.

Only two of the panel's recommendations dealt directly with satellite reconnaissance in the context of "freedom of space." Neither recommendation called for the immediate start of a "crash" program. Rather they indicated the satellites' potential for reconnaissance and the need for a framework to protect them.

The first of these two recommendations responded to legal issues relating to freedom of space. At that time there was no legal definition for where airspace ended and "space" began. Now that launching a satellite into orbit was technically feasible, the question of who controlled space became a central issue. Land's intelligence panel saw a major opportunity for the United States to establish the legal precedent. The TCP report argued that it was necessary to reassess the principles and practices of international law in this area. The departments of Defense, Justice, and the Treasury examined the issues of satellites and international law in the wake of the panel report. They concluded that jurisdiction in space was really not a concern since a satellite in space orbited above the atmosphere and thus would not violate the Civil Aviation Convention of 1944 or international law. By July 1955 the NSC agreed with the panel on the importance of a satellite to test the principle of freedom of space.[19]

The second recommendation also related to freedom of space in the context of overflights. Looking to the long-term goal of space-based reconnaissance, the panel pushed for a small satellite as a stepping stone to future operations. Significantly "intelligence applications warrant an immediate program leading to very small artificial satellites in orbits around the earth. Construction of large surveillance satellites must wait upon adequate solutions to some extraordinary technical problems in the information gathering and reporting system and its power supply, and should wait upon development of the intercontinental ballistic missile rocket propulsion system. The ultimate objective of research and development on the large satellite should be continuous surveillance that is both extensive and selective and that can give fine-scale detail sufficient for the identification of objects (airplanes, trains, buildings) on the ground."[20] Establishing a legal precedent in space, the small satellite would help to protect future reconnaissance satellites. Thus overflight would be legal in space, preventing the Soviet Union or other nations from attempting to stop such activity. The call for a satellite was ahead of its time, as the relevant technology had not yet matured. Certainly satellite reconnaissance fit with the TCP report's call for use of the latest technology in gathering intelligence.

No other open recommendations of the intelligence panel emphasized specific solutions beyond the use of science and technology to create a U.S. advantage. The emphasis on satellites takes on greater significance in light of a secret meeting of Eisenhower, Killian, and Land in November 1954.

Toward the U-2 and Satellite Reconnaissance

Arthur Lundahl, the only CIA official who grasped the concept of increasing data collection, reinforced the idea of overhead reconnaissance. As the head of the CIA photographic interpretation group, he played an important role in the interpretation of many photographs, including the product of the U-2 aircraft when it became available. In August 1954 Trevor Gardner of the air force and Philip Strong of the CIA discovered plans for a revolutionary high-altitude reconnaissance plane, designed by Kelly Johnson, that eventually became the U-2. The air force had shown no interest in the aircraft. In mid-August Strong passed his knowledge of the plane on to Land, who expressed great surprise that the air force had rejected what appeared to be one of the most promising intelligence platforms available. He took the information to James Baker, who had been trying to figure out what type of aircraft would be necessary for overhead reconnaissance. He found the U-2 design ideal for the task. Thus began a long-term relationship between Edwin Land and the U-2.[21]

Although it would become an extraordinary aircraft, the U-2 could not be a permanent solution for all intelligence needs. To do its job the U-2 had to violate international law by penetrating another nation's airspace, which could have severe political repercussions. A more permanent, and politically more feasible, answer was necessary. The TCP report's identification of the key problems relating to satellites—the information-gathering/forwarding system, power supply, and booster rocket—shows that Land and his panel were informed about satellites and their technical problems.

We do not know exactly how much the panel knew and, more important, how much it told Eisenhower. By November 1954 the entire panel found the intelligence situation so depressing that its members needed to speak with the president. Later that month they met with him secretly in the Oval Office, probably on the 24th. No official record of this secret conclave exists (Eisenhower often had such gatherings in the White House); most of the information about it comes from memoirs.[22] According to secondary sources, the major participants, along with Land and Killian, were Secretary of State John Foster Dulles, CIA Director Allen Dulles, Secretary of Defense Charles Wilson, Air Force Secretary Harold Talbott, and air force generals Nathan Twining and Donald Putt.[23] The president's appointment schedule for this date lists an off-the-record meeting at 8:15 that morning but does not show Killian and Land as attending. A memorandum by Andrew Goodpaster about the U-2 decisions of that day indicates that this was the gathering.[24]

This secret meeting has raised a great deal of speculation over the years. While what exactly was discussed at the meeting will likely never be known, it is clear from the remaining evidence that Land and Killian presented two highly compartmentalized pieces of information to the president. This material became part of a secret annex to the TCP report. It was hand-delivered to the president in February 1955 after the formal submission of the TCP report. The annex itself has since disappeared. According to R. Cargill Hall, official historian for the National Reconnaissance Office, the annex did exist but has since either been lost or destroyed— probably the latter.[25]

One part of the annex dealt with the future Polaris ballistic missile system, the other with overhead reconnaissance and a radical new aircraft design that Land and Killian felt was an essential source of intelligence.[26] This aircraft, the CL-282 (later the U-2), seemed to offer a short-term solution to the intelligence problem, helping with defense planning, assessing the Soviet bomber threat, and, in the long run, warning of an attack. Eisenhower authorized its development and gave the CIA responsibility for its operation. The overall cost was probably going to be in the range of $35 million, with both the CIA and the air force providing the funds. Eisenhower realized that since its use would violate international law, its *other* costs would be far higher should it become public knowledge. The secretary of state also anticipated difficulties but argued, "We could survive through them."[27]

Years later Killian indicated that the participants discussed satellites, but as a long-term goal for intelligence.[28] Unfortunately, no memorandum of the meeting dealing with satellites has survived. Eisenhower's penchant for security was very effective in this case. Killian's memoirs describe the meeting and the TCP report, which came out on February 14, 1955. He praises Eisenhower for his willingness to listen to radical proposals and act on them. He also mentions an important point: the most significant and sensitive recommendations on overhead reconnaissance appeared not in the report but in the annex. Worrying that information on the U-2 might leak out, the president ordered its omission, except for the special appendix that was for his eyes only. Killian also states that the report left out a specific recommendation on satellites. Both he and Land accepted that omission. Like Assistant Secretary of Defense for Research and Development Donald Quarles, they saw satellites as a long-term answer, not an immediate possibility.[29]

Eisenhower's decision on November 24, 1954, to go ahead with the U-2 and eventually with satellite reconnaissance was a unique and difficult one. Recognizing that U-2 overflights would violate international law, he showed by his actions his commitment to closing the "intelligence gap" that obscured American knowledge of the Soviet Union. The equally challenging decision about an expensive satellite program, even though its legality and its technical feasibility were still questionable, was clearly visionary. The official General Operational Requirement (March 1955) for this satellite program followed the TCP report by a month and indicated the speed at which the president committed himself. These were not the decisions of a hands-off chief executive.

Eisenhower's "Open Skies" Proposal (March–August 1955)

Realizing that the long-term viability of such systems would necessitate government support, the administration took steps to provide as much legal protection for the U-2 and satellites as possible. To legitimize and protect the U-2 Eisenhower attempted to link photographic overflight with arms-control verification. The closed nature of Soviet society was the major impediment to any form of arms control. There was no practical method short of on-site inspection to verify compliance with agreements, and the Soviets rejected this completely. Grasping the opportunity Eisenhower proposed the use of politically sanctioned aircraft overflights as a means of circumventing Soviet hostility to inspection. This proposal would provide legal protection for the U-2 program then under development and allow both superpowers to halt the escalating arms race.[30]

The "Open Skies" initiative represented Eisenhower's attempt to legalize overflight—an innovative effort to regain the initiative in the cold war and move toward arms control. The concept emerged from a study group that met in Quantico, Virginia. With Nelson Rockefeller (special assistant to the president) as chair, the group consisted of Deputy Secretary of Defense Robert B. Anderson, Adm. Arthur Radford, Harold Stassen (the president's special assistant on disarmament), and eight other people. It convened in June 1955 to evaluate the possible role of aerial reconnaissance in achieving substantial disarmament; its members were unaware of plans for the U-2 or for a satellite system.[31] The core of the proposal was a U.S.-Soviet exchange of data on military capabilities, organization, and the location of units. Mutual overflight and aerial reconnaissance would then verify and monitor this information. In other words, this exchange would help to establish a base-line understanding of both countries' order of battle. Overhead reconnaissance could then monitor any arms agreements to ensure compliance. This arrangement would reduce the risk of surprise attack, ease tensions, and increase trust. To expedite this plan both countries would make facilities available for landing and refueling planes, and both sides would agree not to interfere with each other's aircraft directly or to conceal activities below them.[32]

Eisenhower could justify the proposal in two ways. First, it was a grand and dramatic gesture that promised big propaganda dividends. At the height of the cold war he could show that he was seeking some means of impressing on the Soviet Union his resolve to find a way of ending the arms race. His scheme was unlike previous proposals for arms verification: it demonstrated American willingness to be innovative in this area. It also captured people's imagination by seemingly extending an olive branch to establish peace. Second, it shrewdly attempted to establish legal precedent and legitimacy for overhead reconnaissance; if it accomplished that, then the U-2 and its successors would not be in danger, and in the long run U.S. security would greatly increase.[33]

Whether or not the Soviet Union accepted the Open Skies proposal was of little concern to Eisenhower. Even before making the proposal, he had committed his administration to developing both the U-2 and satellite reconnaissance. The proposal was a clever attempt to gain Soviet acceptance of American overhead reconnaissance. It gave the United States the moral high ground, while providing a rationale for something that Eisenhower was going to do anyway. For him Soviet acceptance was not a requisite. The United States had nothing to lose and much to gain both militarily and politically from the proposal. The Soviets, however, had everything to lose and little to gain if overhead reconnaissance became standard practice. The Kremlin relied on a closed society to conceal its weakness and ensure the illusion of strength. Therefore most people in the Eisenhower administration expected Soviet rejection almost from the start.[34]

Over John Foster Dulles's strong opposition, Eisenhower presented the Open Skies proposal on July 21, 1955, at the Geneva Summit.[35] In the middle of his speech on disarmament in which he asserted, "No sound and reliable agreement can be made unless it is completely covered by an inspection and reporting system adequate to support every portion of the agreement," he removed his glasses, turned to the Soviet delegation, and spoke:

> Gentlemen, since I have been working on this memorandum to present to this Conference, I have been searching my heart and mind for something that I could say here that could convince everyone of the great sincerity of the United States in approaching this problem of disarmament.
>
> I should address myself for a moment principally to the Delegates from the Soviet Union. . . . I propose, therefore, that we take a practical step, that we begin an arrangement, very quickly, as between ourselves—immediately. These steps would include:
>
> To give each other a complete blueprint of our military establishments, from beginning to end, from one end of our countries to the other; lay out the establishments and provide the blueprints to each other. Next to provide within our countries facilities for aerial photography to the other country . . . ample facilities for aerial reconnaissance, where you can make all the pictures you choose and take them to your own country to study, you to provide exactly the same facilities for us . . . and by this step to convince the world that we are providing as between ourselves against the possibility of great surprise attack, thus lessening danger and relaxing tension.[36]

Although politely agreeing that Open Skies had merit, Soviet leaders never seriously considered it.[37] In one sense, however, the proposal was successful: it paved the way for future acceptance of overhead reconnaissance by establishing the link between over-flight and verification of arms control. Eventually this evolved into the term "national technical means of verification" in Article 5 of the SALT treaty of 1972.[38]

The IGY Satellite (February–May 1955)

Having failed to provide legal safeguards for U-2 aerial surveillance by the summer of 1955, Eisenhower turned to protecting satellite reconnaissance. In the wake of the TCP report he had immediately asked all the relevant branches of government to review and comment on the areas of the document relating to their spheres of interest; he had asked the State Department to comment on international law issues.[39] Before these studies could bear fruit, however, other events accelerated satellite development.

The TCP report emphasized the need to determine legally where airspace ended and outer space began.[40] Scientific and psychological issues aside, it was the potential legal ramifications of satellite reconnaissance that most concerned Land and his panel. The mere idea of orbiting a satellite had already sparked the interest of other people within the administration. Indeed the call for a scientific satellite antedated the TCP report. In the fall of 1954 the worldwide scientific community called for a satellite to support scientific research. As a result in October 1954 the Comité Special Année Géophysique Internationale (CSAGI) 1957–58 (the Special Committee for the International Geophysical Year) invited all participating nations to launch a satellite during the IGY. The U.S. National Committee for the IGY concurred and on March 14, 1955, recommended that the United States begin such a program.[41]

Concern over the future of "international" space led Deputy Secretary Donald Quarles to lead the drive to send up an American satellite to mark the IGY. Quarles was an unusual supporter of the IGY satellite concept. Not a believer in untried technology, he doubted that a reconnaissance satellite would be available in the near future. It was not until 1958, when the American IGY satellite got off the ground, that he and others finally realized that a spy satellite was possible. Long before that, however, Quarles was aware of satellites' potential value for intelligence gathering and of air force initiatives in this regard and the resulting need to establish legal precedent. In early 1955 he quietly pushed the military to develop a scientific satellite, which in the long term would assist in establishing the legal precedent for freedom of space.[42] He did this at two levels. Internally, within the military, he began investigating possible avenues for fulfilling the TCP report and IGY calls for a satellite by initiating a study of military satellite options. Externally he supported the American IGY committee's calls for a satellite program, initially through correspondence and later by pushing space policy at the NSC level.

Quarles's Quest for Military Support

Quarles's internal quest for a satellite program began on April 1, 1955, with an inquiry into the value of such a program. He asked the Department of Defense's Research and Development Coordinating Committee on General Sciences (CGS) to review the military's research and development plans for satellites and submit a recommendation on action.[43] Robert W. Cairns, chair of the CGS, replied on May 4. Limiting the scope of its review to research and development on scientific satellites, the CGS found many benefits in the development of both a small, inert, trackable satellite and a satellite with scientific instruments and recommended establishment of a satellite program within the Department of Defense. The main concern was one of scale.

A small, inert satellite (weighing between five and ten pounds) would probably supply a great deal of information about the upper-atmosphere conditions through which satellites, and missile warheads, would have to travel. Ground-based observation of the vehicle would obtain the data. A satellite with scientific instruments would yield a great deal more information, including a range of details about orbital space that was currently unknown. It could monitor conditions such as temperature, available solar energy, solar radio noise, meteor collision, and cloud patterns. Valuable insights would emerge from tracking small satellites, from information relating to gravitational variation and ion content in the atmosphere, and from better understanding of how satellite orbits would shift because of the earth's oblateness and rotation. All these elements would of course have direct implications for any broader space program.[44]

The CGS noted four possible military options for a satellite program. The first, which the navy initially proposed on March 23, 1955, was for a triservice satellite project called ORBITER. The navy envisioned using the REDSTONE missile as a primary booster to deploy a five- to ten-pound satellite with an elliptical orbit ranging from two hundred to eight hundred miles. The outlook was quite discouraging; probably only about one in three attempts would succeed. To maximize the chances the navy argued for four attempts. By adding additional stages, the rocket could increase the system's payload and reliability. Estimates for the cost of ORBITER ran at about $5.5 million plus logistical expenses (and an additional $3 million for instrumentation), and launch would occur probably in autumn 1957.[45]

Second, on April 15, 1955, the navy suggested a satellite configuration that had a VIKING missile as a first stage. Using fewer motors than ORBITER, this prototype would probably have a more reliable propulsion system and hence slightly better performance. The navy confidently predicted that such a missile could lift a ten-pound satellite into a 334-mile-high orbit, and a larger satellite (roughly forty pounds) was possible, if a lower orbit at around 200 miles was acceptable. Overall cost for this program (which the navy saw as a backup and second-generation system to ORBITER) would be roughly $7.5 million.[46]

Third, the air force proposed a satellite closely related to the WS-117L satellite concept. It would use an ATLAS ICBM as a primary booster and an AEROBEE-HI rocket as a second stage. Because of ATLAS's higher thrust potential, the air force predicted that its design could lift a satellite payload of one hundred to two hundred pounds into orbit. Expecting to be able to orbit the satellite by the end of 1958, the air force estimated the program's cost at $50 million to $100 million. The air force's final proposal involved the CONVAIR test missile (then part of the ICBM program) as a booster to lift a satellite weighing about two thousand pounds into orbit. However, the incorporation of a satellite role for this system seemed to the air force neither necessary nor desirable because the satellite would slow ICBM development.[47]

While acknowledging that the Department of Defense probably possessed the technical capability to launch a scientific satellite during the IGY period of July 1957 to December 1958, the CGS's review concluded that the many complex technical problems meant that there was no guarantee that the satellite would be ready in time. As well, any program had to be part of continuing research and development, which would produce a series of improved satellites. Unfortunately the CGS did not recommend any one program. Noting the prestige and value of satellites as a basic research tool, it argued that Defense should pursue development of all three (ORBITER, VIKING, and ATLAS-AEROBEE) to make success more likely and provide a range of capability.[48]

Quarles's Quest for Nonmilitary Support

Even as the CGS was preparing this report, Quarles was mustering nonmilitary support for the IGY program. Aware of the TCP report's call for a satellite to support freedom of space, Quarles went to the U.S. National Committee for the IGY in February and March 1955 to gain its backing for a satellite program. He prompted Alan Waterman, director of the National Science Foundation, to contact both the State Department and the CIA concerning the U.S. IGY committee's request for such an effort.[49] State responded favorably in discussions on March 22 and in a memorandum of April 27. Robert Murphy, a deputy undersecretary at State, thought the program in the national interest. However, he felt that State would be unable to judge whether a satellite was strategically feasible and so deferred to the military on this issue. More important, he argued that "arrangements among the pertinent government agencies should go forward, with the understanding that the U.S. representatives of CSAGI will report on our intentions in the IGY at the appropriate time."[50]

CIA Director Allen Dulles and his deputy, Richard Bissell, were also responsive to the concept of the IGY satellite. Both thought expediency was a necessity and suggested two approaches. The proposal could go forward from the Department of Defense or it could be directly presented to the NSC. To accelerate the process Dulles agreed to take up the matter with the NSC's Operations Coordinating Board, but Waterman got the impression from both State and the CIA that it was up to either him or Quarles to initiate formal action. Quarles, however, decided to act on his inside knowledge. Quietly he pushed the United States to develop a scientific satellite that would work toward establishing legal precedent. He expected the U.S. IGY committee to formally request development of a scientific satellite, which it did on May 18, 1955.[51]

NSC 5520 (May 1955)

Keen to see a national space policy to support satellite reconnaissance, two days later, on May 20, 1955, Quarles submitted a draft statement on a policy to the NSC—the first articulation of a national space policy. It related not to the two military programs but to the value of a satellite launched under IGY auspices. Quarles's policy paper was similar to the TCP report's recommendation about establishing the legal basis for freedom of space.[52]

In the "Draft Statement of Policy on U.S. Scientific Satellite Program" (NSC 5520), Quarles confirmed the TCP's findings that the United States had the technology to produce a satellite of five to ten pounds, and, more important, it would be able to orbit one in 1957–58. The draft also stated that since the TCP report had appeared the Soviet Union had apparently taken the initiative. On April 15, 1955, Quarles announced that the Astronomic Council of the Soviet Academy of Sciences had established a permanent high-level "interdepartmental commission for interplanetary communications."[53] There were also significant indications that the Soviet Union might be working on a satellite of its own. Quarles went on in NSC 5520 to identify some of the advantages of a satellite, which included the gathering of scientific data, psychological benefits, and increased prestige from being first in space. Among the most valuable benefits was the opportunity to test freedom of space. Preliminary studies by the executive branch uncovered no legal obstacles under international law to orbiting such a satellite. Therefore the IGY could establish the legal precedent of overflight by a satellite. Based on these factors NSC 5520 called for government support of the IGY satellite.[54] Arguing that a satellite was not a militarily offensive weapon, the paper suggested that the benefits of orbiting a satellite would greatly outweigh the $20 million that the program might cost. NSC 5520 expected the military to take the lead in launching the satellites, as it had the most experience with rockets. Emphasizing the peaceful nature of American space efforts, in conjunction with the IGY commitment, it claimed that those factors would ease the possible backlash over a satellite launch.[55]

Quarles was not alone in comprehending the value of the scientific satellite. The CIA also backed the notion in its discussion of the TCP report. Asserting that "successful launching of the first satellite will undoubtedly be an event comparable to the first successful release of nuclear energy," the CIA argued that the first nation into space would gain incalculable prestige. More significant, the satellite would be a stepping stone toward a larger satellite capable of providing early warning and reconnaissance of the Soviet Union.[56] The potential psychological impact of being first in space and the military intelligence advantages of a satellite impressed Nelson Rockefeller, who firmly backed the idea. Aware that Soviet propaganda would criticize a U.S. satellite, he emphasized the value of IGY auspices for the effort.[57]

NSC 5520 committed the United States to launch a civilian satellite during the IGY period. Eisenhower hoped that a civilian program would resolve the legal issue and minimize the risk of an international dispute. The IGY's blessing gave it additional legitimacy and made strong Soviet opposition to overflight less likely. Despite the psychological and prestige effects of a successful launch, at least one member of the NSC grasped its more subtle implications. During the discussion of NSC 5520, Allen Dulles observed "that it was very important to make this attempt"—the IGY satellite was crucial for the future of satellite reconnaissance.[58]

VANGUARD (July 1955–June 1958)

Although the United States could have waited for one of the international scientific meetings in August 1955 to announce its IGY program, the administration was increasingly worried that the

Soviet Union could tell the world about its own satellite program at any time. So it seized the initiative, announcing its satellite in a press statement on July 29, 1955. Careful wording stressed its scientific benefit for everyone concerned. The press release stated, "This program will for the first time in history enable scientists throughout the world to make sustained observations in the regions beyond the earth's atmosphere."[59] Press Secretary Jim Hagerty pointed out that the president was pleased that American scientists would help colleagues from all nations to benefit from this project.[60] The announcement was also the first step in securing legal protection for satellite reconnaissance.

The scientific satellite program, VANGUARD, proved more of a challenge than anyone expected after its inauguration on September 23, 1955.[61] Due to Eisenhower's desire to separate the ICBM program from civilian space activities, the scientific program was only loosely placed under naval management. Initial financial projections were overly modest, and a small workforce that had to do everything from scratch quickly produced cost overruns.[62] Politically the decision to create a civilian program was sound and stemmed from the emphasis on the endeavor's scientific and nonmilitary nature; a launch on a military missile would have sent the wrong message. Also at risk was the secrecy of American missile technology when the scientific data became public. Furthermore the ICBM program could not afford a delay for a satellite shot. Accordingly the president kept the two programs separate.[63]

Air force historian Lee Bowen argues that a lack of monetary support by the secretary of defense intensified financial problems. Money for the program arrived too sporadically to support rapid development. Following the launch of *Sputnik*, difficulties with finances and the army's claim that it could have launched a satellite far quicker led to attacks on the program and the president. Eventually the army's Jupiter C experimental booster helped to orbit the first American satellite, EXPLORER, on January 31, 1958.[64]

Concerns over possible conflict between the ICBM efforts and VANGUARD were troubling. Less than a year after the public announcement of VANGUARD, the NSC learned about this problem. In early May 1956, during an NSC meeting, Secretary of

Defense Wilson recommended that to prevent conflict between the two programs VANGUARD should be given lower priority that the ICBM efforts. This seems to support Bowen's assessment of Wilson's support for VANGUARD. The financial problems were already evident, and the meeting revealed that costs were rapidly escalating. The National Science Foundation had asked for an additional $28 million for more satellites, and Secretary of the Treasury Humphrey warned the NSC that the projected costs for the VANGUARD program were approaching $60 million to $90 million, a far cry from the original estimates. The president then reminded Humphrey that the higher figure was for the proposed doubling of the number of satellites to launch. This would be reviewed as needed, but Eisenhower did not expect the price to exceed $60 million.[65]

The hemorrhage of funds in the VANGUARD program did not stop at a mere $60 million. By the end of January 1957 the cost had risen from the original $20 million estimate to $83 million, while the experts expected the launch of a satellite only in October 1957. The National Science Foundation sought an additional $30 million to increase the number of satellites, raising total expenditures to $113 million, something both Eisenhower and the Department of Defense were unwilling to countenance. This pattern repeated itself in May 1957, when cost overruns indicated that the anticipated total would reach $110 million.[66] Expensive technical problems had slowed development of a reliable booster. VANGUARD was not just attempting to establish a legal precedent; it was also trying to achieve a major technological breakthrough.

In the meantime the legal questions that NSC 5520 and VANGUARD were to solve were proving a major stumbling point for satellite reconnaissance. In the wake of the TCP report the State Department had begun to reassess the legal ramifications of satellite overflight, believing that no state could claim territorial sovereignty at such altitudes. It expected that space would be open to all countries so long as those seeking to use it could surmount the technical problems.[67]

Eisenhower hoped that the scientific satellite would define space as international, just as the earth's oceans are free domain outside the territorial limits of nations. This "great common" in space would allow open access and prevent interference. To achieve this he emphasized VANGUARD's scientific and peaceful purpose. Once the Soviets accepted the principle of overflight, reconnaissance would receive at least tacit legal sanction, which would help to prevent countermeasures against satellites. For both Eisenhower and Quarles the Soviet launch of *Sputnik* in October 1957 solved the problem to some extent. Quarles suggested that the Soviet Union had in fact "done us a good turn, unintentionally, in establishing the concept of freedom of international space."[68] It was not until the mid-1960s, however, that the United Nations was able to develop the rudiments of a legal regime for space. Initially satellite reconnaissance existed in quasi-legal limbo. The Soviet Union stopped protesting about U.S. space espionage once its space capabilities matched those of its great rival.[69]

NSC 5814 (August 18, 1958)

After *Sputnik* U.S. space policy adapted quickly. In June 1958 the NSC reviewed a new policy on outer space in NSC 5814, which offered a more comprehensive approach than NSC 5520. NSC 5814 systematically assessed the value of American security in space, reconsidering the issue of the legal boundaries of outer space. For example, a theoretical upper limit of air space might be the highest altitude at which an aircraft could maintain enough lift for flight and/or the lowest orbit a satellite could maintain without encountering atmospheric drag.[70]

Officially NSC 5814 emphasized the American desire for peaceful use of outer space, but the document shows that the United States considered some military uses of space essential for national security. The most important of these involved reconnaissance satellites to gather military and other intelligence on the Soviet Union and to detect possible missile launches against the United States. The political challenge was to create a legal position that allowed certain satellite activities, such as reconnaissance, without creating excessive political repercussions. Reconnaissance satellites seemed a possible means of implementing the Open Skies proposal that the Soviets had rejected in 1955.[71]

The TCP report was a pivotal moment, when Eisenhower solicited advice from sources outside the national security bureaucracy. Their expertise could both invigorate defense efforts and challenge the conventional wisdom, especially that of the military. Land's intelligence panel, which the Science Advisory Committee of the Office of Defense Management created ostensibly to advise the president on issues relating to possible surprise attacks, in fact had a far greater impact. It dramatically clarified the military and civilian need for intelligence and drew the chief executive into the process. Advocating both the U-2 and satellite reconnaissance as sources of intelligence, the panel, with Eisenhower's full support, changed the course of the cold war.

The TCP report linked Eisenhower's principal challenges: ensuring national security through intelligence and strategic warning, defending economic security, and blunting the military's opposition to sound fiscal planning. Good intelligence would, in theory, allow the government to balance defense spending by silencing the shrill calls of the military and its supporters for larger budgets. More significant, overhead reconnaissance might monitor Soviet military activity, verify arms-control agreements, and in the long term strongly counterbalance undue military influence over the political leadership.

We can see how important these goals were to Eisenhower by the effort he was willing to put into achieving them. He realized that overhead reconnaissance as an intelligence source would work only within a legally benign international system. The U-2 aircraft had to violate international law in order to do its job, but reconnaissance satellites had no legal status because of the ambiguity of space law. Eisenhower used the power of his office to protect the U-2 program as much as possible. He probably knew that the Soviets would never accept Open Skies. He made it as a legitimate offer, however, because it had the potential to protect aerial reconnaissance. He may also have theorized that it would put the United States on the moral high ground by linking photographic reconnaissance with the peaceful goal of verifying arms control.

With satellite reconnaissance Eisenhower likewise tried to protect a future intelligence asset by securing legal protections. U.S.

acceptance of the IGY satellite proposal and NSC 5520, even though it would cost millions of dollars (something Secretary Humphrey repeatedly railed against), shows how highly Eisenhower valued the potential for this form of intelligence collection. The IGY satellite was an attempt to use a scientific satellite to achieve a military/intelligence goal; it would establish the legal precedent that could then somewhat justify spy satellites.[72]

VANGUARD thus has to be seen as more than the American contribution to the IGY efforts. It was also part of Eisenhower's attempts to create legal protection of space-based reconnaissance. By gaining worldwide acceptance of satellite overflight, no nation could legitimately protest military satellites. VANGUARD, in essence, was running interference for WS-117L. The fact that the Soviets orbited the first satellite only made it easier for Eisenhower. Once they had established the legal principle of freedom of space, they would be unable to backtrack by protesting an American satellite. From 1958 on, U.S. space policy shifted to block military uses of space while supporting peaceful ones. Of course it defined reconnaissance satellites, which provided stability and protection from surprise attack, as "peaceful."

Individually and collectively the Technical Capabilities Panel, the IGY satellite, and VANGUARD were seminal to Eisenhower's strategy for national security in the cold war. They charted a cohesive, long-term effort to safeguard the future of U.S. intelligence gathering. They show the extent of the president's gamble, applying innovative methods to solve the problems that he faced. Yet in 1958 the very satellite that he was trying to safeguard had yet to get off the ground.

PART 2

WS-117L

[4]

Origins

RAND and Satellite Reconnaissance (1945–54)

Despite years of research on the subject, the space issue never
reached the upper levels of the Truman White House. There was
no Truman space policy, and space issues remained largely the
realm of a small group of engineers and analysts.

—Dwayne Day

In an era dominated by the doctrine of massive retaliation, cities
were the main targets and bombers were the main threat. For such
a military outlook, reconnaissance from space represented a useful
but scarcely essential capability.

—Robert Perry

By 1954 a variety of forces were converging within the Eisenhower
administration to produce a drive for satellite reconnaissance. Con-
current with the growing awareness of the value of intelligence
and rising political will to take action, a small group of civilian
consultants and military officers was looking for solutions. These
experts were seeking to conceptualize a new reconnaissance system—a process that began in 1946—under the auspices of the
RAND Corporation.[1]

RAND played a threefold role. First and foremost, after it became
clear in 1946 that satellites were technically feasible, RAND examined the elements necessary for a launch, such as stabilization and
power supplies and possible political and psychological repercussions. Second, by 1950 RAND had helped the air force (which
Truman created in 1947) through its early jurisdictional battles to
emerge as the dominant agency for satellite development. Third,
its studies helped to define a military use for satellites.

This chapter first explains the context and content of RAND's report of 1946 on the feasibility of satellites and shows how the air force employed the results to achieve dominance in the satellite turf wars of 1945–50. It then examines RAND's work on satellite reconnaissance that led to its groundbreaking 1951 study and looks at the aftermath of that report. Finally it considers RAND's FEEDBACK project of 1953–54, which dealt in detail with technical issues concerning a reconnaissance satellite. As I discuss in chapters 5 and 6, a number of key players in RAND's preparation of its crucial reports of 1951 and 1954 later played major roles in WS-117L.

RAND and the Emergence of the Air Force (1945–1950)

By the end of 1945 the U.S. Army had already launched a concerted effort to dominate missile projects, using captured German scientists and missiles, creating the White Sands Missile Range in New Mexico, and funding several rocket-related programs. On October 3, 1945, the Navy Department expanded its Bureau of Aeronautics (a research-focused bureau) to include the Committee for Evaluating the Possibility of Space Rocketry.[2] It did so at the prompting of Cmdr. Harvey Hall, special scientific assistant to the head of the Radar Section in the Bureau of Aeronautics. The most important of its many specified goals was conducting feasibility studies for employing an artificial satellite to relay naval communications. Within a month the committee agreed that it was technically possible to launch a satellite into orbit, and a contract went to California Institute of Technology's Jet Propulsion Laboratory (JPL) for it to undertake further study. In 1945, however, the engineering and preliminary work alone looked likely to cost in the range of $5 million to $8 million. In the face of financial cutbacks and demobilization, there were no funds for a project that lacked direct military application.[3]

In an effort to preserve and expand on the existing work, the navy turned to the U.S. Army Air Forces (USAAF) to try to pool resources for a program. On March 7, 1946, it suggested the formation of a satellite committee to the members of the USAAF on the joint-service Aeronautical Board. The government had formed the board during the First World War to review new developments in aeronautics. Consisting of high-ranking members from the army and naval air arms, this board met monthly and acted as a bridge between the two services on air matters. In 1946 the Joint Research and Development Board took over some roles from the Aeronautical Board, which was finally disbanded in 1948.[4] The new body's members agreed in March 1946 that the potential of satellites justified further discussion. The army representatives to the Aeronautical Board's Research and Development Committee agreed to investigate the "extent of Army interest by discussions with [Major] General [C. E.] LeMay [director of research and development]."[5]

According to R. Cargill Hall, this consensus did not last long. Whereas most accounts portray the USAAF as delaying a decision on a joint program until after it could consult LeMay, pushing back the meeting from April 9 to May 14, 1946, Hall maintains that the USAAF rejected the joint effort prior to the April 9 meeting because it believed that missiles were in its bailiwick. Interservice rivalry quickly ended any hopes of such a collaborative effort.[6]

At issue was the question of roles and missions, problems that plagued the military for years. The army saw the navy's proposal as a possible threat to its own rocket research, Project HERMES, which was to develop and test missile technologies using the German V-2, with General Electric holding the contract.[7] The army saw development of rockets as an extension of the artillery's role on the battlefield, so it regarded the navy's proposal as poaching on what it considered an army mission. To forestall naval encroachment, the office of the USAAF's commanding general took the position that its branch had to demonstrate its competence and interest in space research to prevent an interservice conference from outflanking it.[8]

Immediately after the decision to defer discussion of the navy's idea, LeMay called for a comprehensive study to establish the USAAF's competence in space matters. A tough-minded officer whose combat experience was in long-range strategic bombing, LeMay realized that simply turning down the navy's joint project would not end that service's interest or efforts. The navy had already contracted several feasibility studies with JPL and several private companies. To prevent further naval encroachment on army air force turf, LeMay realized that he needed scientific studies to demonstrate the air force's interest and dominance in the field. He turned to Douglas Aircraft's Research and Development project, or Project RAND, for a scientific report on the feasibility of orbiting a satellite. Douglas had set up Project RAND in March 1946 to continue the close cooperation between the military and the scientific community that had begun during the war. Realizing that science and technology had transformed warfare, Gen. H. A. P. Arnold and several key individuals from the aerospace industry tried to keep the relationship alive, and RAND was the result.[9]

LeMay needed the report from RAND in time for the meeting of the Aeronautical Board's Research and Development Committee on May 14, 1946. RAND put in a rush effort, delivering its preliminary report on May 2 and a revised copy on May 12.[10] Although not directly focused on reconnaissance, RAND's first report, "Preliminary Design for an Experimental Earth Circling Spaceship" (1946), proved seminal. First, it established that it was technically feasible to orbit a satellite. Second, it showed that any satellite would have significant repercussions. The RAND scientists expected that such a vehicle would become one of the most potent scientific instruments of the twentieth century. Third and more significant, an American device would impress the world with U.S. technology and have a psychological impact on other nations.[11]

The document provided an in-depth analysis of the possibility of building and launching a satellite. Discussing various configurations for the rocket booster, it theorized about the possibility of sending up a payload of roughly five hundred pounds to an altitude of about three hundred miles. Such a satellite could stay in orbit for about ten days to gather scientific data. The report envisioned a multistage rocket (with either two or four stages, depending on the fuel) and analyzed the conditions in which the satellite would operate, the problems of attitude control, the threat from meteors, orbital trajectories, and the possibility of recovery. It determined moreover that it was theoretically possible to control the forces involved sufficiently to place a human being into orbit. The overall cost to orbit the first satellite would be perhaps $150 million over five years.[12]

This pivotal work linked the scientific and military importance of both rockets and satellites. The only difference between a satellite booster and an ICBM was that a satellite required slightly more energy for orbit. Thus developments for one would aid the progress of the other. Predicting (correctly, as it turned out) that the missile would become the primary delivery vehicle for nuclear weapons, the document suggested several roles for a satellite, the most important being reconnaissance.[13] The value of the satellite was very clear: "It should also be remarked that the satellite offers an observation aircraft which can not be brought down by an enemy who has not mastered similar techniques. In fact, simple computation from the radar equation shows that such a satellite is virtually undetectable from the ground by means of present-day radar. Perhaps the two most important classes of observation which can be made from such a satellite are the spotting of the points of impact of bombs launched by us, and the observation of weather conditions over enemy territory."[14] Support missions, such as attack assessment and communications, were also feasible for satellites. As early as 1946 scientists had proposed observing the earth from space, and the report's demonstration that a satellite was theoretically and technically feasible set the stage for RAND to look carefully at questions of utility.

LeMay's attempt to use the RAND report to demonstrate the army air force's interest and competence in space was partially successful. It showed that the USAAF was looking at the problems and potential of satellites and effectively blocked the navy's proposal for a joint effort. It did not, however, prove the army air force's dominance in space matters; rather it muddied the waters. During its May meeting the Aeronautical Board's Research and Development Committee did not assign responsibility for developing a satellite to either contender. Instead it forwarded a summary to the Aeronautical Board, which, with half its members from the army and half from the navy, deferred judgment until higher authorities made a decision.[15]

The issue of institutional control of satellites remained unresolved, and in January 1947 Rear Adm. Leslie Stevens (assistant chief, research and development, in the Navy Bureau of Aeronautics) went over the heads of the Aeronautical Board's committee directly to the Joint Research and Development Board (JRDB).[16] He requested the formation of an ad hoc committee to coordinate all phases of an Earth satellite program and assign the control over such efforts to one of the services. In essence, as one historian of the pre-NRO satellite program, Robert Perry, maintains, the navy hoped to circumvent the army air force's dominance in the satellite field. Unfortunately the JRDB was a coordinating body with no authority to decide policy or act. To make matters worse, by trying to circumvent the Aeronautical Board and its committee, the navy only raised the ire of its competitors. According to Hall, the JRDB took note of Stevens's ideas and remanded them back to the Aeronautical Board for review before a final decision.[17]

Events soon overtook the process. Congress created the National Security Act, which President Truman signed into law on July 26, 1947, and which reorganized the players in the early satellite drama. One of the act's major elements was the reorganization of the military in the light of new technology. It separated out the old army air forces as the core of the new independent air force. With the emergence of the independent air force in 1947, the two-way war over satellites became a slugging match among the three services. To be sure, the satellite issue was only a small part of the interservice rivalry. The three branches were fighting not just for control of the latest technology but for their share of everything from defense spending to personnel, industrial support, and popular opinion. For the air force this rivalry took on greater importance. Seeing itself as the first line of defense because of its nuclear-strike role, the air force thought of itself as the dominant American striking arm, relegating the other services to secondary roles. Yet this put the air force in a difficult position. Needing to keep its bomber and fighter forces strong enough to act as a deterrent, it also had to counter other services' perceived attempts to undercut its roles. Being the youngest service, it was extremely sensitive about its independence. Thus for the air force interservice rivalry for cash, resources, and popular opinion quickly became a fight for its survival as an independent military arm.[18]

The Research and Development Board (RDB) absorbed the JRDB on September 30, 1947, and became part of the new Department of Defense. The RDB was to assist the secretary of defense in coordinating the armed forces' research and development activities and answered to him; Vannevar Bush was its chair. No longer a coordinating organization, the RDB could now create policy and exercise some direct control over the making of decisions. Aware of the security implications of satellites and the need for an effective policy, the RDB directed its Committee on Guided Missiles to coordinate satellite policy, which removed the roadblock at the Aeronautical Board. The committee assigned its Technical Evaluation Group to examine the issues and assign responsibility.[19] The RDB killed the navy's plans to orbit a satellite because their proposed program lacked immediate military utility. In 1946–47 cuts in the defense budget severely curtailed or canceled many programs. A highly speculative and risky program that did not fill a pressing military need did not stand much chance. The only activity that the RDB sanctioned was continued study of satellite utility through RAND.[20]

The former army air forces had been more successful. Although the first RAND study did not create much enthusiasm for a satellite program, it did demonstrate the program's feasibility. Unable to obtain control over satellite efforts, the army air forces were able to gain the Aeronautical Board's approval to continue examining space projects (and reaffirm it later with the JRDB). It ordered a second RAND study, which over the winter of 1946–47 produced a series of documents on various aspects of satellite flight that defined the problems and benefits of a project. These papers triggered no immediate action but laid the groundwork for future planning. Following the creation of the air force on September 18, 1947, they received further consideration. On September 25 the air force ordered its Air Material Command to evaluate all the RAND studies from the perspective of technical and operational feasibility.[21]

The Air Material Command's response was conservative, and

its report certified technical feasibility but cautioned that the complexity and cost might be untenable. The paper questioned the air force's ability to maintain funding for such large-scale research and development. Noting that the practicality of this kind of system was in question, it urged a study to determine air force requirements, functions, and scheduling that would permit development at a later date. In the meantime guided missiles had priority.[22] Satellites remained on the back burner, although RAND continued doing relatively cheap and useful feasibility studies.

Although the air force did not pursue a satellite program immediately, it decided to assert its supremacy in space research. Lt. Gen. H. A. Craig, deputy chief of staff for material, concluded that, despite current financial limitations, progress in research on guided missiles meant that a full-scale satellite program would become feasible in time. On January 12, 1948, Craig urged Gen. Hoyt S. Vandenberg (vice chief of staff) to assert the air force's responsibility for space, thus assuring its dominance. Vandenberg agreed and three days later signed a formal policy document—the first clear declaration by any service regarding control of a space program.[23]

Vandenberg's "Statement of Policy for a Satellite Vehicle" affirmed the air force's claim on space research and development policy: "The USAF, as the service dealing primarily with air weapons, especially strategic, has logical responsibility for the satellite." Indicating that it was making rapid progress with guided missiles, he linked satellites with the missile sciences: "The problem will be continually studied with a view to keeping an optimum design abreast of the art, to determine the military worth of the vehicle—considering its utility and probable cost—to insure development in critical components, if indicated, and to recommend the initiation of the development phase of the project at the proper time."[24] This statement put the other services on notice that the air force would study the matter of orbiting a satellite to assess its military value and costs.

The air force's director of research and development authorized its Wright Field agency near Dayton, Ohio, to put Vandenberg's ideas into effect. It fell to RAND to monitor the state of technology and inform the air force of technical developments. In February 1948 Brig. Gen. Alden R. Crawford (Wright Field) instructed RAND to develop components and techniques for an eventual launch, thereby maintaining the effort at a study level only. RAND in turn subcontracted important research to aerospace companies such as North American and RCA and to institutions of higher learning such as Boston University. It helped determine the feasibility and effectiveness of complex satellites, and by 1950 research was focusing on two future aspects of the space program. These illustrated changes within the RAND community that had not yet reached the military or the U.S. administration. RAND had begun to examine satellites' psychological/political implications and, more important, their military utility.[25]

Satellite Reconnaissance and the RAND Report (1950–53)

The shift in emphasis from feasibility to utility helped RAND convince the air force's Directorate of Intelligence to undertake further research into satellites. Within RAND satellites had strong support from many people, including Merton Davies and Amrom Katz, who became advocates, monitors, and even guiding voices of "expert knowledge when it came to satellite reconnaissance from 1954 to 1960." These enthusiasts transcended their role as objective providers of knowledge to become partisan advocates.[26]

The first step was to determine the psychological and political ramifications of orbiting a scientific or a military satellite. In a study that appeared in early October 1950, Paul Kecskameti directly addressed these issues. While examining them from the perspective of the United States being first into space, he focused on the impact this would have on the Soviet Union and world opinion. In many regards his report was prophetic.[27]

He believed that a satellite, whether scientific or military, would have significant political ramifications. The worldwide reaction depended on how the United States was viewed. The technological achievement would inevitably increase perceptions of American strength and technical and scientific superiority. Rival or hostile states would perceive a satellite as a threat to their interests. The Soviet Union in particular would see it as a challenge and a demonstration of its own weakness. Its propaganda would emphasize the American threat. Friendly nations would see the success as an affirmation of American scientific, technical, and military prowess and thus find it reassuring.[28]

Kecskameti's interpretation is entirely logical and represents a clear understanding of the likely situation. To counter these problems he advised the United States to neither conceal the launch nor play up its military value. To maximize its political and psychological value, the U.S. government had to keep the public abreast of its development, while emphasizing its peaceful and scientific roles. This approach would prevent the illusion of a nefarious military purpose. The alternative approach of total secrecy and then a sudden release of information would have a greater psychological impact. Unfortunately this was likely to push even neutral countries toward hostility. In the long run, no matter what approach the United States followed, hostile states would still react negatively to an American satellite. Although advance publicity would blunt some of the dramatic effect, it would be better to diminish the effect rather than risk a strong backlash. In either case the Soviet Union, which depended on secrecy to conceal its military weakness, would react harshly and equate an American satellite with reconnaissance and perceive it as a direct threat.[29]

Finally Kecskameti raised concern about the legality of overflight. There was no legal precedent for a satellite flying over a sovereign nation. Without resolution of these issues, such a use of space could lead to serious international legal repercussions. Kecskameti was accurate with his prediction that a satellite was "bound to be a spectacular event, causing a worldwide sensation." The post-*Sputnik* reaction certainly lived up to his expectations, complete with "estimates of aggressive intent behind the development and use of the instrument."[30] His only incorrect prediction was that the United States would orbit a satellite first.

Kecskameti's report shows that RAND clearly anticipated the benefits and risks of satellite reconnaissance. It also set the stage for later actions by the Eisenhower administration to establish the legal framework for both satellites and overhead reconnaissance, but it would not lead to a satellite program. Before that could begin the air force had to demonstrate a compelling military need—only that would win funding. Beginning with their initial study, RAND scientists had looked at a variety of roles for a satellite, among which one crucial mission stood out: reconnaissance. But before a satellite for that purpose could emerge, RAND had to prove its technical feasibility in gathering useful intelligence and work out a viable design. By 1954 it had produced both. This achievement coincided with the Eisenhower administration's desire for better surveillance of the Soviet Union and its conclusion that a reconnaissance satellite might be the right vehicle.

The idea of overhead reconnaissance was not new in 1950; it was as old as aviation itself. By 1945 aerial photography allowed for extremely high-resolution color photographs. Researchers, including Dr. James Baker, had developed camera lenses with focal lengths as large as 240 inches,[31] which automatically compensated for alterations in air pressure as an aircraft's altitude changed. By war's end photographic equipment had become very precise, and the interpretation of photographic images had become much more technically sophisticated. Army air force doctrine, however, considered overhead strategic reconnaissance only in terms of specific functions in wartime: for identifying and locating targets vital for aerial attack, for their defenses, and for assessing bomb damage.[32] The concept of preconflict reconnaissance to gather general intelligence simply did not exist within the U.S. military.

The advent of atomic weapons had forced the United States to reassess intelligence and strategic reconnaissance. Breaking away from the narrow military view, many began to equate "pre-D-day," or peacetime, reconnaissance with an ability to warn of surprise attack. In the wake of Pearl Harbor and the dawning of the nuclear era, peacetime reconnaissance became more important. The first clear call for a new type of reconnaissance came from Gen. Henry H. "Hap" Arnold (USAAF) in November 1945. Like Eisenhower, he understood that atomic weapons made the nation's wartime experience obsolete. While Pearl Harbor had been costly, the United States was able to recover and rebuild its military to wage a global war; but an atomic Pearl Harbor would be quite different, threatening the American heartland itself. The general warned Secretary of War Robert Patterson that if U.S. leaders were to prevent atomic attack they needed "continuous knowledge of potential enemies." This included information not just on troop deployments but on economic, industrial, political, and scientific developments. Arnold was unwilling or unable to say how to collect such data; the man who could—and did—was a junior officer, air force lieutenant colonel Richard S. Leghorn.[33]

Leghorn was a graduate of the Massachusetts Institute of Technology with a bachelor's degree in physics. At MIT he became friends with James R. Killian Jr. Leghorn's career encompassed military and civilian occupations that helped to shape his views on reconnaissance and science. Following graduation in 1939, he worked for Eastman Kodak before joining the army reserves as a lieutenant in the Ordnance Corps. In 1940, following a meeting with Maj. George W. Goddard, commander of the army air corps' Aeronautical Photographic Laboratory at Wright Field, Leghorn transferred to the corps and began work in aerial photography. As a member of Goddard's team, he searched for methods to glean more data from photographs. He also met several major figures in reconnaissance circles who came to play a major role in satellite reconnaissance, including Amrom Katz (a civilian physicist and future RAND scientist), Lt. Walter J. Levison (a physicist), and astronomer James G. Baker, who designed camera lenses both during and after the war.[34]

Leghorn reunited with his Dayton colleagues to provide aerial reconnaissance during the CROSSROADS nuclear testing at Bikini Atoll in October 1945. The military needed trained crews, and Leghorn became deputy to Col. Paul T. Cullen, commander of the photographic unit. During the journey to the test site Leghorn had access to the *United States Strategic Bombing Survey (Europe)*, which assessed the effectiveness of bombing operations in Europe. The report noted that a lack of strategic reconnaissance before the war and during the initial stages of the bombing had hampered targeting and wasted effort. Leghorn studied the conclusions, specifically the call to improve coordination among intelligence services in collecting and evaluating information. The document also encouraged involvement of civilian scientists in intelligence gathering and concluded that "the combination of the atomic bomb with remote-control projectiles of ocean-spanning range stands as a possibility which is awesome and frightful to contemplate."[35]

The Bikini testing in July 1946 impressed on Leghorn the case for prehostility reconnaissance. For the first time he saw the result of an atomic blast. Although there is no evidence that he knew about General Arnold's comments on the importance of intelligence, he shrewdly reached the same conclusions. Reliable peacetime intelligence would decrease the threat of a nuclear Pearl Harbor. Leghorn quickly realized that regular overhead strategic observation would solve the intelligence dilemma. He also decided that a specialized aircraft would reduce the chance of detection. He repeatedly shared his views on overhead reconnaissance with his colleagues from Wright Field.[36]

Among the early converts was the physicist Duncan Macdonald, head of Boston University's Optical Research Laboratory (BUORL). Macdonald thought highly of Leghorn's ideas and invited him to speak at BUORL's dedication on December 13, 1946. Leghorn's comments were the first public discussion of peacetime overhead reconnaissance. He argued that atomic weapons made peacetime aerial reconnaissance imperative. Nuclear war's destructiveness made surviving a surprise assault and mounting a successful counterattack unlikely. Only information from aerial reconnaissance could prevent such a surprise.[37]

With the scars of Pearl Harbor still fresh in the American psyche, the consequences of a lack of vigilance were clear. Aerial reconnaissance of the sea approaches to Hawaii could have provided some warning. Now the constant threat of total destruction from a single attack created a pressing need for intelligence. The Pearl Harbor effect was dramatic in the military: it led to calls for larger budgets, more striking power, and a stronger defensive stance. However, this outlook prepared the United States for a possible confrontation but provided no warning. Leghorn called for something different: the ability to warn of an attack before it happened.[38]

Leghorn envisioned a world in which most intelligence came from overhead reconnaissance in daylight and, if the technology allowed, at night as well. He believed that this was the only way to penetrate a totalitarian regime keen to preserve its secrets. He admitted that this activity was illegal under international law but insisted that it was essential for national security. Thus a method to minimize the chance of detection was necessary. He envisioned an aircraft that flew at extreme altitudes, *not* satellites. At this point in time spy satellites were still a speculative idea.[39]

Following the nuclear testing Leghorn returned to Eastman Kodak, but he did not stay out of military intelligence circles for long. The army air forces started overflights of the Soviet Union as early as 1946, when they employed aircraft—mostly RB-29s (modified Boeing B-29 Superfortresses)—patrolling the Soviet periphery to collect intelligence. Carrying cameras of limited focal length (thirty-six inches or less), they flew within a few miles of Soviet territory taking oblique photographs. The army air forces were especially curious about the areas around the Kola Peninsula (especially the port of Murmansk), the area north of Leningrad, and the Chukotskiy Peninsula (across the Bering Strait from Alaska). They saw these regions as potential staging points for any attack on the United States. Since the planes had to stay outside Soviet territory and had only cameras with short focal length, the pictures were of limited value. Hence these flights concentrated on gathering electronic intelligence. Despite gradual improvements to the cameras, oblique photography never provided the depth of coverage that the United States required.[40]

In the wake of the first Soviet atomic bomb test in 1949, U.S. restrictions on the penetration of Soviet airspace were quickly pushed aside. As the cold war chilled East-West relations, the air force (especially the Strategic Air Command, the nuclear striking force) argued for penetrating Soviet air space to gather intelligence. With evidence of Soviet nuclear testing in hand, General LeMay (then commander of SAC) recommended to President Truman that the United States begin overflights of the Soviet Union to collect evidence of possible preparations for a surprise attack. Still using converted bombers, these missions inflamed emotions in the Soviet Union and were very risky in the face of its air defenses. Penetrating its airspace posed two problems. First, the planes' limited range meant that there were vast areas that they could not observe. Second, they were extremely vulnerable to enemy fire, so each flight increased the chances of an international incident. From 1952 to 1955 seven incidents did occur, in which thirty crewmen were either killed or missing in action.[41]

These problems forced the air force to find another method of obtaining intelligence. By 1955 it had tried a variety of schemes, the most innovative being Project GOPHER, which unsuccessfully released hot-air balloons carrying cameras across the Soviet Union. But GOPHER and the development of the U-2 aircraft during 1955–56 were only stopgap solutions.[42]

By the time Kecskameti's 1950 report appeared, RAND was already on the way to solving this problem: using satellites for reconnaissance. This work took the original RAND studies on orbiting a satellite and married them to the concept of overhead reconnaissance that Leghorn advanced in 1946. Culminating in two separate studies, the idea of overhead reconnaissance evolved by 1954 into a clearly defined and articulated call for a satellite to provide vital intelligence for the next fifty years. By November 1950 RAND's thinking had advanced to that point, and it recommended to air force headquarters that it extend research into some aspects of a proposed reconnaissance system.[43]

Following a meeting with RAND, Col. Bernard Schriever requested that the air force's Directorate of Intelligence provide specific intelligence requirements for a reconnaissance satellite. The directorate's reply on March 17, 1951, identified high-quality images as the primary requirement. The proposed satellite would need picture resolution sharp enough that experts could identify harbors, oil storage areas, large residential and industrial targets, and airfields and collect weather data. The images had to provide enough clarity to allow for correction of maps and charts and be able to cover the entire Soviet Union in a period of weeks, provide continuous daytime coverage, and record the information in more or less permanent format. Representatives from the directorate visited RAND on March 2, 1951. The scientists simulated a satellite photograph by taking a photo of Los Angeles and relaying it via television to Mount Wilson, California, and back before photographing the monitor screen. Photo interpreters examined the photo and agreed that it would satisfy the minimum requirements. Maj. Gen. C. P. Cabell, the author of the memo from the Directorate of Intelligence, recommended starting the reconnaissance satellite project "with a view toward the present urgent need for such a reconnaissance system, rather than a future need."[44] The demonstration that a satellite could provide the necessary type of intelligence was only the beginning.

RAND's study, "Utility of a Satellite Vehicle for Reconnaissance" (April 1951), examined five different areas of importance relating to satellite reconnaissance. Starting with a discussion of orbits and the ground area covered, the report went on to cover the reconnaissance process, control of the satellite, power supplies, and reliability of a robotic system. It analyzed comprehensively the capabilities of satellites for reconnaissance, paying particular attention to the use of television, communications, and the problem of the electrical power supply. These were all limiting factors of the satellite.[45]

The report found that the optimum orbital altitude for a spy satellite was about 350 miles, with an inclination of roughly 53 degrees from the equator. This meant that the satellite would orbit the earth fifteen times per day. Due to the earth's rotation, the satellite's orbital path over the main areas of interest would shift by about eight hundred miles with each orbit. Thus every eighty-seven or so days the satellite would return to its original flight path, having covered the entire area of interest. Unfortunately, because of the earth's rotation, daylight coverage was possible only during alternating thirty-five-day periods. Thus it would take two satellites to give adequate coverage.[46]

This altitude seemed to be the best balance between the forces that limited the satellite. Higher orbits promised the vehicle longer life and more time to convey information but needed larger booster rockets to reach orbit; either image quality would be poorer because of the greater distance, or a larger payload would be necessary to allow for bigger camera components. Lower orbits meant smaller boosters and higher image resolution but shorter life because of friction in the denser atmosphere. Friction slows a satellite, causing it to ride lower in the atmosphere, and generates heat. Eventually the satellite will reenter the atmosphere and burn up. An altitude of roughly three hundred miles seemed likely to make for a satellite life span of two years.[47]

The RAND scientists felt that the ICBM program provided the obvious answer as to how to launch the satellite. An ICBM was a suitable primary booster if high accuracy was possible in attitude control. This was essential to ensure that the camera system faced the earth properly at all times. The scientists at RAND discussed several options, including the use of gyros, flywheel stabilization, and attitude sensing, but presented no definitive solution.[48]

Power consumption was the main limiting factor for the system. The satellite needed enough power to run equipment for an extended period of time. This included not only the camera and attitude control but also the radio, ground communications, storage of ground commands, and timers. J. E. Lipp and the other RAND scientists predicted that the satellite would need about five hundred watts of power for up to a year, while staying within strict weight requirements. With a 1,000-pound payload, only 250 pounds could go to produce electricity. Conventional means of producing power were ruled out because of the huge amount of fuel they would need. Therefore the RAND report proposed a nuclear reactor employing radioactive isotopes to produce heat to generate electricity.[49]

The study team now tried to figure out how to retrieve data based on the predicted orbit pattern. It quickly dismissed the idea of recovering the intelligence through the use of a reentry capsule. Just as with the ICBM program, any attempt at physically recovering anything from space was seen as impossible at that time. The extreme heat of reentry was expected to destroy the capsule. While the USAF was working on solving this problem for the ICBM program, the technology had not yet been found to allow a successful reentry. Until the technology caught up to the idea, alternative methods of information recovery were needed. Two alternative methods for gathering photographic intelligence were examined in detail: photostatic facsimile transmission and the adaption of standard television technology.[50]

Photostatic facsimile transmission was a reliable technology then in use. It used standard camera film to record images. These images were then electronically scanned, converting them into electrical impulses in a method similar to the standard wire photo of the time. A radio link could then relay this information back to Earth. While tried and tested already on Earth, the main limit of the system was that it required a lot of film. A month's operation was expected to consume about three-fourths of a ton of film. Weight wise, this was cost prohibitive unless reusable film became available.[51]

The alternative option was to adapt current television technology. While only allowing daylight observation, the scientists felt that it would work for pioneering reconnaissance efforts. Basing their predictions on current technology, RAND scientists predicted photographic resolution of roughly two hundred feet.[52] Though suitable for weather data, such low resolution was not good enough for surface surveillance. It would detect major airfields, highways, railways, and large factories but not much more. Productive intelligence needed fifty-foot resolution. The study also predicted that assessing bomb damage would necessitate resolution of about ten feet. Increasing resolution by increasing lens magnification meant either decreasing the area covered per photo (so total coverage would take longer) or technically improving the camera system. The scientists were confident that resolution of forty feet would eventually be feasible.[53]

The study's final issue was reliability. A satellite in space would have to operate independently and perform perfectly for a year or more in a vacuum and at extreme temperatures. In addition it had to survive the excessive vibration of the launch, motion changes, and gravitational forces. None of these expectations seemed unrealizable. The scientists believed that technology "on the shelf" could sustain at least thirty-five days of operations. Through research and development, this might be extended to a year.[54]

In conclusion the 1951 RAND study assessed available technology and related it to aerial reconnaissance. It also helped to prove that a military satellite was feasible and made it clear that any problems with television reconnaissance—the most likely system—would relate to engineering. RAND saw the basic science as already extant.[55]

The RAND scientists disseminated and promoted their work at various levels of the military. After release of the report, military representatives and members of BUORL received briefings at Wright Field. In the audience was Amrom Katz, chief physicist at the Air Force Reconnaissance Lab and one of the greatest skeptics about satellite reconnaissance. The briefing by Lipp and his colleagues generated a mixed response. Some listeners found the findings encouraging, except for the expected poor resolution of satellite images. The air force photo interpreters thought the resolution inadequate for their purposes, which seriously challenged the RAND team. Katz and three of the BUORL participants used the resolution problem to discount the whole proposal.[56]

Katz, Walter Levison (assistant director, BUORL), Duncan Macdonald (director, BUORL), and Col. Richard W. Philbrick (air force liaison officer to BUORL) found a test to refute RAND's proposals. The test was conducted at Wright Field in November 1951. It used an 8-mm camera and coarse film to take reconnaissance photos from an altitude of thirty thousand feet to mimic the satellite's performance. Katz and his colleagues were stunned by the results. Expecting to see nothing of value, they were able to clearly identify roads and bridges in the Dayton area, the airfield itself, and several landmarks. Katz became an instant convert, joining RAND in 1954 and pushing for satellite development.[57]

During the spring of 1951 three developments occurred at roughly the same time as RAND's report. The first was the activation of the air force's Air Research and Development Command (ARDC) in April 1951. Working with the deputy chief of staff for development, the ARDC was to improve research and development within the air force. Since war's end that service had concentrated on building up its combat arms, but this effort had failed to provide for future needs. The ARDC began to play a greater role as the satellite program moved toward development.[58]

Second, the USAF recalled Colonel Leghorn to active duty in April 1951 because of the Korean War. Initially stationed at Wright Field, he was reassigned to the weapon-system procurement office as chief of reconnaissance systems. There he surveyed the requirements for reconnaissance and assessed possible appropriate systems. Initially he worked in coordination with Colonel Schriever, assistant for Air Force Development Planning (AFDAP) in Washington. Schriever was preparing Development Planning Objectives (DPOs) for the service's roles in areas such as tactical air operations. In August 1951 Leghorn was transferred to Washington under Schriever's command to help prepare a new DPO for intelligence and reconnaissance and to act as the air force's liaison officer with RAND.[59]

Leghorn was in a unique position in this assignment. He worked with some of the most influential people in Washington: Edwin Land, Carl Overhage (chief of Eastman Kodak's color laboratory), James Baker, Edward Purcell, and Burton Klein from the RAND Corporation. These men shared their views on intelligence gathering. Curtis Peebles maintains that Leghorn's liaison duties also put him in contact with Merton Davies and the RAND satellite proposals.[60] Thus Leghorn was at the focal point, both within and outside the Pentagon, for much innovative thinking.

Leghorn further refined his views on intelligence in 1951–52 to help the "counter force" strategy for nuclear weapons that he had begun to advocate. Such a strategy calls for targeting of a potential enemy's nuclear arsenal with your own. In the event of war your nuclear strike would then attempt to disarm your enemy. Leghorn realized that intelligence was the key here and that the military would need prior intelligence of Soviet strategic assets. As Leghorn put it, "Our qualitative intelligence and reconnaissance capabilities constitute the primary problems, and without extraordinary action, these might delay adoption at operational planning levels of strategies with emphasis on counter force operations."[61]

Overhead reconnaissance was the only practical method of gathering this intelligence. Leghorn emphasized use of an unmanned aircraft with a credible cover story, such as pursuit of scientific or weather data. Recognizing that manned aircraft would have to suffice, he thought them vulnerable and useful only over areas with weak air defenses.[62] How much he knew about RAND's satellite proposals is in question. According to Davies, Leghorn's contacts at RAND and conferences on intelligence gathering exposed him to the idea of satellites. Davies claims that he had convinced Leghorn about reconnaissance satellites and their inclusion within the framework of air force needs; however, Leghorn considered these long-term goals.

A third development in 1951 coincided with release of the RAND report, emergence of the ARDC, and Leghorn's return. An agreement between MIT and the air force to study defense issues led to the creation of the Beacon Hill Study Group under the auspices of Project Lincoln. Schriever's AFDAP branch had called for such a body to consider U.S. intelligence problems and to suggest better ways of gathering and processing intelligence. These civilian and military experts worked from July 1951 until completion of their report the following June.[63]

The Beacon Hill experts included Baker, Killian, Land, Leghorn, and Overhage. We can see Leghorn's influence in one of the report's crucial recommendations: "We have now reached a period in history when our peacetime knowledge of the capabilities, activities, and dispositions of a potentially hostile nation is such as to demand that we supplement it with the maximum amount of information obtainable through aerial reconnaissance. To avoid political involvement, such aerial reconnaissance must be conducted either from vehicles flying in friendly airspace, or—a decision on this point permitting—from vehicles whose performance is such that they can operate in Soviet airspace with greatly reduced chances of detection or interception."[64] The report recommended improvements in sensors and development of aircraft and other means of flying over the Soviet Union. Borrowing from Leghorn's ideas (as they appeared in his DPO), its writers emphasized vehicles capable of peacetime reconnaissance with a minimal risk of detection and preferred unmanned flights for penetration of another nation's airspace. They outlined several options for reconnaissance, including high-altitude balloons (this became Project GOPHER in the mid-1950s), higher-altitude aircraft, and either the Snark or the Navaho air-breathing missile. The group considered aircraft overflight politically risky except for lightly defended areas. Whatever means the military selected, it needed a credible cover story if diplomatic complications occurred.[65]

The Beacon Hill Group did not ignore the RAND satellite proposals. According to Davies and Peebles, Leghorn eventually became a convert to satellite reconnaissance. But the Beacon Hill Group received no direct RAND input and invited no RAND personnel to speak, let alone to participate on the steering committee. Several RAND personnel did attend meetings and had contact with members outside these sessions, but they were there as guests.

The Beacon Hill report (June 1952) assumed that satellites were possible in the long term. Believing that making satellite reconnaissance a reality in a timely fashion would be very expensive, the group concluded that it would be better to apply such funds to developing other, more promising technologies. Both Land and Leghorn were strong supporters of balloon and aircraft reconnaissance and advocated these options instead.[66]

Leghorn's role in the development of overhead reconnaissance was played out in two final acts. First, in November 1952 he briefed the Air Force Air Council, the CIA, and the NSC on reconnaissance requirements and progress to date on efforts to meet these needs. Second, he submitted his DPO to Schriever in January 1953, after which he returned to Eastman Kodak before eventually starting the Itek Corporation, which produced payloads for the future CORONA program. In this DPO he argued that high-altitude reconnaissance was the best means of gaining intelligence from the Soviet interior. He forcefully supported development of high-altitude balloons and specialty aircraft for overhead reconnaissance, with balloons providing area coverage and planes penetrating areas of interest that photo interpreters deemed worthy of closer observation. Satellites would play a role in the *distant* future. Until then the endeavor required a specially built aircraft capable of operating at over seventy thousand feet. In the wake of Leghorn's DPO, the air force began looking at designs of high-altitude aircraft for strategic reconnaissance, and the winner was the U-2.[67]

Within a few months of the Beacon Hill report's completion in June 1952, Americans elected Eisenhower as president. For such a cost-conscious, intelligence-aware chief executive, the report became more than just an air force document. As a foundational work its conclusions were picked up in the even more important TCP report of 1955. The Beacon Hill paper reinforced the value of intelligence and of overhead reconnaissance (the very foundation of the RAND study of April 1951). More important, its emphasis on an unmanned vehicle with a small likelihood of detection and interception seemed to fit satellite reconnaissance. The problem lay not in convincing experts of feasibility but in doing things quickly.

Despite the best intentions to expedite satellite development, inertia within the air force meant progress was slow and tedious. Even as the Beacon Hill experts were considering satellites, some air force officers saw development as premature. They suggested instead a further feasibility study. Thus on December 19, 1951, the air force authorized RAND to make recommendations on development of reconnaissance satellites.[68] This slow movement frustrated the RAND experts. The old notion that reconnaissance was a wartime operation was hard to kill. Ironically the one agency that should have been advocating peacetime photographic reconnaissance, the CIA, showed no interest. It still focused on human intelligence, which emphasized penetration agents and clandestine collection. Its very small photographic-interpretation unit had no real authority.[69]

Despite air force and CIA resistance, the ideas from RAND, Leghorn, and the Beacon Hill group were slowly filtering up through the air force chain of command. In December 1952 Lt. Gen. Thomas D. White (deputy chief of staff for operations) promoted the idea of satellite reconnaissance. Citing the Soviet hydrogen bomb, he argued that Washington desperately needed to "obtain reconnaissance and surveillance data leading to knowledge of Soviet capabilities and intentions before the beginning of hostilities." White mentioned a high-flying aircraft but thought a satellite reconnaissance vehicle the best solution. Invoking RAND's predictions for resolution and satellite specifications, as well as political and psychological values, he recommended that the air force immediately issue a requirement for a satellite.[70]

Project FEEDBACK and Fine Details (June 1953–March 1954)

Despite frustration RAND returned to the drawing board for yet another study, called Project FEEDBACK. This effort was aimed not only at justifying a satellite system but at specifically defining the components required. To support this effort, in 1952 RAND signed contracts with a number of companies to examine particular aspects of a reconnaissance satellite. These contracts included several with RCA to look at adapting television cameras, radiation-recording devices, and other equipment, and North American Aviation, which studied orbital sensing and control systems. The air force also supported these studies. In July 1953 the Communication and Navigation Laboratory at the Wright Air Development Center in Dayton contracted with North American Aviation for a study of pre-orbital guidance systems. The air force even arranged for the Atomic Energy Commission to begin work on small reactors suitable for a satellite. The government funded most of these projects under its existing contract with RAND through a special supplement for fiscal year 1953 specifically for the satellite research.[71]

By this time the air force was ready to take a more active role in satellite reconnaissance. In May 1953 it ordered the ARDC to take over the FEEDBACK program by June 1 and to investigate the feasibility of starting to develop auxiliary nuclear power plants for satellites. This led to a series of meetings between the ARDC and RAND. The ARDC found RAND's efforts to date impressive and quickly grasped the importance of satellite reconnaissance. In September, with ARDC support, RAND pushed the air force to authorize contracts for system design within a year and then begin full-scale development.[72]

In the waning months of 1953 pressure for satellite development was building. Seeking to pull together elements of the endeavor for better management, the ARDC gave the program its first official designation. Identified as Project 409-40, "Satellite Component Study," the program was also given an unofficial project number (WS-117L) for later system development. The Wright Air Development Command took responsibility for the program and began work to demonstrate the feasibility of major satellite elements—specifically television components, attitude and guidance control, and auxiliary power-plant subsystems. In January 1954 the project received the unclassified title "Advanced Reconnaissance System" and an engineering-project designation, MX-2226.[73] Lacking only official authorization WS-117L was ready for development.

In March 1954 the long-awaited, two-volume FEEDBACK study finally appeared. It was released just as the Defense Department was reporting that ICBMs were technically feasible and at about the same time as the Killian Commission was working on the TCP report. Unfortunately, very little detail on the report has found its way into the historiography because it has only recently been declassified in its entirety. Portions became available in the 1960s for Robert Perry's work "Origins of the USAF Space Program," but they indicated only some of the RAND study's conclusions. Most early historians refer to the FEEDBACK report but give only vague indications about a reconnaissance satellite using television technology. More recent work by Jeffrey Richelson, R. Cargill Hall, and William E. Burrows provides more details about image resolution and the satellite's longer-term implications. They maintain that the report's description of a satellite capable of continuous Soviet surveillance spurred the creation of a system for satellite reconnaissance.[74] Unfortunately, these sources do not convey either the report's scope or its impact on the decision to develop satellite reconnaissance. To better understand the report and its significance to the air force, it needs to be examined at some length.

The FEEDBACK report, like RAND's April 1951 studies on satellite reconnaissance, represented a refinement of not only the engineering data but also the fiscal and political aspects of such a system. It was the culmination of work begun in 1946, in the sense that many of the same scientists who had been advocating satellites then played a role in FEEDBACK. As such it represented the collective wisdom of some of the top minds in the field. In the report RAND formally recommended that the air force undertake the development of a satellite vehicle at the earliest possible opportunity. It felt that this decision ought to be made at a high policymaking level and that both the decision and the program itself should be covered in a blanket of secrecy. Predicting that the process would take about seven years at a cost approaching $165 million, the experts cautioned that this estimate could double or triple depending on a variety of circumstances.[75]

To help explain the satellite system that FEEDBACK proposed, specific aspects of the program need to be examined. The report recommended a two-stage launching rocket to put the vehicle into orbit. With an overall length of about eighty-one feet and a total takeoff weight of about 178,000 pounds, the rocket would use two main boost engines and two gimbaled motors to generate a takeoff thrust of 285,000 pounds. The gimballing of the smaller engines was part of the rocket's control system. Small motors linked these engines and the control systems, allowing the rocket to adjust the exhaust and thus change direction. Approximately 80 percent of each stage went for storage of the fuel and oxidizer (in this case, gasoline and oxygen). The second stage was the satellite itself, which consisted of the reconnaissance payload, weighing 1,500 pounds (about one-third of the second stage's total dry weight), related items (recorder-playback systems and so on), guidance and attitude control, transmitters and receivers, and the power plant. The camera, antennae, and horizon scanners would be mounted at the rear of the satellite, above the motor for the second stage.[76]

The only prerequisite for a launch site would be proximity to the latitudes from which the military wanted intelligence. Thus for surveying the Soviet Union a location near White Sands, New Mexico, or Patrick Air Force Base on the California coast would be acceptable. The actual ascent would follow a pattern that has become almost the norm: after a vertical launch, the two stages would be fired in sequence to achieve sufficient altitude. At that point, using an inertial guidance system, the rocket would tip over to the correct angle for the desired orbital path before coasting into a generally circular orbit. The second stage would ignite again briefly for a final application of power to create a stable orbit. The inertial guidance system acted as a point of reference to measure satellite motion to maintain course and attitude. To stabilize the platform a system of gyros, accelerometers, analog computers, and servo controls measured and compensated for any pitch, yaw, or roll. A horizon scanner would keep the camera aiming at the earth. The operational life was likely to be at least one year.[77]

The power source for the satellite was to be a water-moderated nuclear reactor designed to heat mercury into a gas that turned a turbine to generate electricity. The reactor itself was a sphere two feet in diameter that would produce about eighty kilowatts of heat energy. Radiators along the satellite's hull would dissipate excess heat and cool the mercury. The RAND experts chose a nuclear reactor due to the need for a long-term power source. Chemical power sources and the radio-isotope-heated power plant from the 1951 report seemed inadequate, and using conventional fuels and oxidizers was impractical due to the massive amounts needed. Solar energy was also rejected due to the primitive state of the technology at the time. The only limit on nuclear reactors, however, was the need to radiate excess heat outside the spacecraft. Because of the satellite's size the experts concluded that such a device would generate about one kilowatt of electricity; more power would require diversion of more of the satellite's weight to the reactor. The overall weight of a plant producing a single kilowatt of electricity would be roughly 450 pounds. For an additional fifty pounds of satellite weight, the reactor could produce two kilowatts. The plans called for a power source of one to two kilowatts.[78]

The satellite that FEEDBACK described was to be part of an integrated system for obtaining photographic intelligence. The report also dealt with testing, ground control, data handling, and other elements. Ground stations would download imagery, aid in tracking and locating the satellite in orbit, relay command instructions to the vehicle, and monitor satellite systems. They would relay intelligence data to a central intelligence center, which would interpret photos, compile them into mosaics, and store them. A two-way radio link would handle communications. The onboard television would relay the image to a receiver at the ground station, whence a transmitter would instruct the satellite. All communications between ground stations and satellites would use microwave-transmission frequencies, which require line of sight for communication. Scientists anticipated that this would reduce jamming and decrease the risk of interception by the Soviet Union. Ideally the satellite would be able to store information before transmitting it to ground stations within the continental United States. Otherwise the ground receiving stations would have to be located closer to the Soviet Union, where they would be under greater security threat.[79]

The satellite's success of course depended on its ability to produce photos with enough resolution for interpretation. RAND understood that initial imagery would at best be the bare minimum acceptable. Conventional aerial photography was using imagery with a map scale down to 1:80,000, the minimum useful for interpretation; resolution of satellite photos was to be only about 1:500,000. Lines of communication and distribution and transportation systems (including highways, railroads, and pipeline and power-line right of ways) would be visible; however, as they appear as long lines they would be mutually indistinguishable, whereas harbors, docks, ships, and related structures would be readily visible. Equally important, repeated observation could also indicate the level of activity in these areas. Photographs with a 1:500,000 scale can reveal airfields, larger military installations, and urban and industrial areas with some degree of definition.[80]

Thus early on experts expected satellites to provide only what the USAF director of intelligence described in 1951 as "pioneer-level" imagery and mapping capabilities, which were adequate and useful for planning bombing operations. Barring weather problems, the system could probably survey the entire Soviet Union in three or so days.[81] Identifying the roles and functions of buildings and other structures would be difficult.

Both RAND and RCA proposed a camera system employing a standard Image Orthicon television camera of the period, attached to somewhat better equipment, which would project the ground image onto two sequentially operating cameras. Because the satellite would operate beyond the line of sight of ground stations, magnetic tape would store images until the satellite was close enough to relay them to the ground station. The state of the art at the time of the report was a video magnetic-tape recorder with a bandwidth of about 1.5 megacycles (Mc). The designers thought that they could increase its speed to 8 Mc while shrinking it to roughly two hundred pounds—a simple engineering problem. The study proposed development of two cameras: a pioneering one, operating at 1:500,000, and one with higher resolution, at 1:125,000, to provide more detailed information on ground targets and wider surveillance coverage.[82]

Employing a scale to indicate an image's quality is not as effective as defining resolution in terms of the size of an object that the camera will see. The 1:500,000 pioneering system would provide ground resolution of perhaps seventy feet in diameter under ideal conditions (if the object was alone in a uniform background), but ideal conditions were extremely unlikely. Therefore RAND and RCA experts expected a two hundred–foot ground resolution with the original system. The camera itself would not move throughout the process. A scanning drum would rotate so that it could record sequentially a strip of ground under the satellite. Each strip would be approximately 374 statute miles wide, with each frame covering roughly 77 square miles. The camera would also compensate for the satellite's forward momentum to minimize blurring of the image. The camera with a scale of 1:125,000 would use similar techniques but with increased resolution.[83]

The satellite system that Project FEEDBACK's two-volume report proposed presented its potential users a groundbreaking concept. Overall it was not exactly what the air force had hoped for, but it was technologically feasible. This document accomplished what the air force had asked RAND to do in 1951. The primary difference between the two reports rests in the depth of detail and documentation. FEEDBACK contained far more technical detail than the 1951 report. Volume 1 dealt very precisely with comparisons between camera systems and justifications for choosing one, details about metals, effects of a vacuum on gasses within the satellite, and so on. It also placed more emphasis on the reconnaissance mission of the satellite. It included samples from a series of test shots that simulated space photos. Volume 2 served in effect as a technical annex. The demonstration of photographic products reflects the strength of the FEEDBACK document, which did not simply recommend satellite reconnaissance but *sold* it. The system could obtain photographic coverage of previously unseen Soviet regions clearly enough to make the enterprise worthwhile.

The overall impact of RAND's efforts from 1946 to 1954 was immeasurable. From 1946 on, RAND scientists were convinced that launching a satellite was technically feasible and would benefit the United States politically, psychologically, and militarily. Every study from 1945 to 1954 strengthened their faith. Colonel

Leghorn's work and his views about prehostility reconnaissance meshed well with emerging views on intelligence. By 1954 the technology was finally catching up to RAND's ideas.

A number of people who worked within RAND on both the 1951 report and FEEDBACK went on to key roles in WS-117L. Other people with whom the RAND group came into contact also had a major impact. Edwin Land, James Killian, Herbert York, and others all became prominent figures in intelligence gathering and within the Eisenhower administration. Just as important, a few air force officers realized not only the feasibility of satellites but their utility. Even before FEEDBACK forward-thinking officers had pushed for satellite reconnaissance. Most writers seem not to have known about RAND's jump-starting satellite development. To quote Bruno W. Augenstein, "The impetus given to satellite work by RAND studies of this era seems mostly forgotten now; but it is doubtful if the program could have obtained a running start without it."[84]

Despite all it achieved RAND could not create the satellite program nor ensure its completion, but it germinated the seed of the satellite effort. The need for intelligence and a president eager to control defense spending allowed the seed to grow. All these forces came together in 1953–54 to produce the first satellite program. The RAND studies were not *the* decisive factor, but they established the necessary specifications for satellites to fly and created a constituency of influential supporters who could sell the feasibility and utility of satellites. The administration and a handful of the air force accepted the need for them. The success of the reconnaissance satellite would depend on research and development by many interested and passionate parties who believed wholeheartedly in the concept.

[5]
WS-117L
Two Stages (1954–57)

By the end of the Eisenhower administration, the foundations of
each of the major military space programs had been laid.
—William J. Durch

The year 1954 saw the convergence of all the forces I have discussed
so far. First, in March 1954 the concept of a satellite for reconnaissance reached fruition in the form of RAND's FEEDBACK satellite
study, which defined the shape of a satellite reconnaissance system and became the foundation for development of the WS-117L
satellite. Second, as RAND was finishing its report, the desire for
military intelligence was leading to creation of the Technological
Capabilities Panel, its two reports, and the U-2. Third, President
Eisenhower began working to protect satellite reconnaissance by
securing legal protection for it through his Open Skies proposal
and the VANGUARD satellite for the International Geophysical
Year. By the end of 1955 he had also accepted the WS-117L program.[1] It fell to the air force to develop satellite reconnaissance.

The WS-117L effort laid the foundation for every military satellite after 1953. Many historians mention it but provide only fragmentary and very incomplete details. For example, the program
received considerable criticism after the success of *Sputnik* in October 1957 because of its slow progress and lack of results. But the
reasons for these delays have eluded most authors, who cannot
isolate and explain the underlying problems.

There are few available sources about development of WS-117L,
especially from 1954 to 1957, partly because of the slow declassifi-

cation of relevant material. Daily logs, correspondence, and even plans and reports are still inaccessible. The only window that currently exists into these years of development comes from two sources: the writings of Robert Perry and James Coolbaugh. Perry, an air force historian, wrote several works, including the multi-volume official history of the satellite reconnaissance program. Despite many blacked-out sources, his history outlines the official documentation that has yet to surface. For insight into daily workings, one may turn to the memoirs of Capt. James S. Coolbaugh, the WS-117L's project officer from December 1953 to March 1957, at the Wright Air Development Center's (WADC) New Development Project Office within the Bombardment Missiles Branch. Coolbaugh joined the WADC in September 1952 when he began to work with the Bombardment Missile Branch under Maj. Sidney Greene, chief of the New Development Project Office. The idea of space-based reconnaissance was not unknown to Coolbaugh. He had knowledge of the RAND efforts on satellite reconnaissance from Maj. Quenten Riepe, the WADC's liaison with RAND, who had kept him up to date on developments.[2] Coolbaugh's memoirs, put together at the request of R. Cargill Hall, are unique as they detail his daily activities as project officer from 1953 to 1957.

The danger of relying on only two sources, one of which is a personal memoir, is that the picture presented of the program will be skewed and incomplete; thus wherever possible, I use alternative sources to verify both Perry's and Coolbaugh's accounts. So although it cannot be considered a definitive account of the early development of WS-117L, this chapter conveys a fascinating glimpse into the development of American satellite reconnaissance prior to *Sputnik*.

As I explained in chapter 4, the Air Research and Development Command's support of the RAND proposals led the air force to merge satellite efforts into a single project. Once work began in January 1954 the system received the unclassified title Advanced Reconnaissance System.[3] Perry maintains that on December 3, 1953, the ARDC ordered the WADC to start work on demonstrating the feasibility of the major satellite components. The WADC assigned the program to the Bombardment Missiles Branch in its Systems

Management Organization.[4] Perry's description of how the WADC was assigned responsibility for the program seems accurate and consistent. However, he does not explain clearly three additional designations: project number 1115; the cover name "Pied Piper," which the press and several authors picked up and took to mean many things; and WS-117L.

Coolbaugh, in contrast, explains all three designations. Project number 1115 and Pied Piper refer to the program's process for bidding on contracts. His wording is very clear: "The evaluation of the three proposals completed the Project 1115, PIED PIPER portion of the satellite program."[5] As I discussed earlier, the project received the name WS-117L unofficially at the end of 1953, but it became official only in 1955, according to Coolbaugh.[6] As 1954 began, then, before RAND's FEEDBACK report appeared, the air force's work on satellites had changed from a semiofficial planning project to a proposed system complete with project number.[7]

Satellite reconnaissance developed in two stages between 1954 and 1957. The first began in December 1953 with Coolbaugh's appointment and ran until roughly October 1955. This period included initial planning for the contract process as well as research on specific components in the WADC's laboratories. In 1955, however, the program transferred from Wright Field to the Western Development Division (WDD).[8] This decision was logical since the satellite's primary booster was likely to be the ATLAS ICBM, which the WDD was developing under Gen. Bernard Schriever. The group at WDD supervised the selection of contracts, further work on components, and contract development. The two stages overlapped, as many personnel who worked at the WADC transferred with the program to the WDD; much of the effort carried over from one period and facility to the other.

First Stage: Wright Air Development Center
(December 1953–October 1955)

In the twenty-two-month initial stage, research and development started from scratch at the Wright Air Development Center. Coolbaugh first obtained a description of RAND's proposal from Major Riepe (the RAND-WADC liaison officer), who was the ideal man to brief him. Intimately aware of the satellite concept, Riepe had turned down Coolbaugh's position as project officer and elected to remain the liaison officer. Riepe's orientation and a RAND briefing on FEEDBACK in mid-January 1954 brought Coolbaugh quickly up to date on the concept for a reconnaissance satellite. Unfortunately the air force had not budgeted for the WS-117L in 1954, so there was no money for research and development and no guarantee about how much funding would be available for the next fiscal year. The lack of funding was the reason Riepe had rejected the position of project officer.[9]

Coolbaugh found that everyone involved in the project at RAND was enthusiastic about the FEEDBACK proposal. He returned to the WADC with helpful information and strong indications of areas requiring immediate action. After discussing the issues with his superior, Major Greene, he put the WADC's laboratories to work on developing components for every major system on the satellite, including a rocket motor for the second stage, an onboard power supply, recording of data via videotape, reliability of the electronics, and atmospheric properties. Coolbaugh was also faced with financial problems and the need to find a system contractor. He makes it clear in his memoirs that he spent a lot of time keeping up with all the work done by the various labs at WADC on the WS-117L project. This led to a very close working relationship with both the on- and off-base labs involved in the program.[10]

Coolbaugh was fortunate that some work had already been started under RAND contract with outside companies. So in some areas he had only to convince the appropriate WADC lab to continue working with the original contractor. This was certainly the case with respect to the horizon scanner for the satellite. One of the most pressing challenges to be overcome was how to stabilize the satellite so that the camera was always pointed at the earth. The solution was the horizon scanner, which, in tandem with control systems, would align the cameras properly. North American Aviation had already started to work on this device. In order to maximize this advantage Coolbaugh approached the WADC's own Communication and Navigation Laboratory to convince it to cooperate with the company to develop and test it. Eager to participate in space research, the lab agreed to contribute expertise and to fund it itself. Because a working horizon scanner was so critical to the success of the WS-117L satellite, Coolbaugh felt that a backup program was needed. So he established a backup program using the Armament Laboratory at the WADC and the Instrumentation Laboratory at MIT, using money left over from other projects. By working on two different designs for the horizon scanner, Coolbaugh felt that he was maximizing his chances for success.[11]

Next Coolbaugh turned to the satellite's need for power. This was a dual problem as the satellite needed both a second-stage engine to establish its orbit and an energy source to perform its tasks. Coolbaugh was fortunate when it came to the rocket engine for the satellite. During discussions with the air force he learned that the Bell Aircraft Company was already developing a rocket motor for the B-58 HUSTLER strategic bomber. With most of the cost for development paid for under its contract for the B-58, all he had to see to was retrofitting the motor to the satellite and related costs. Providing power for the satellite was not as simple. The FEEDBACK report called for a nuclear power plant, but no one had ever conceptualized such a small reactor. The best that Lt. Col. Edward Hall of the Power Plant Laboratory could do was introduce Coolbaugh to people doing research on nuclear-powered aircraft engines. Although they could not solve his problem immediately, they agreed to review RAND's findings and make suggestions.[12]

Related to the power supply were various problems with the satellites' electronics, the most pressing of which was how to record the satellite imagery. This was a vital element since without recorded imagery, the satellite was useless as an intelligence platform. Because such electronics did not form part of the lab structure at the WADC, and RCA had undertaken the original study for RAND, Coolbaugh met with Jim Huckaby, project manager of RCA's video-recording development team. Many technical problems had to be worked out. The prototype system at that time ran standard studio-quality audiotape at very high speeds (360 inches per second) past two recording heads. A seventeen-inch reel of tape could hold only about four minutes of video information. Since the satellite was to operate in an environment almost devoid of gravity, the reels' rapid spinning would create destabilizing forces in the satellite. The stabilization system would be hardpressed to compensate for even the seventeen-inch wheel of tape. Anything larger, and the vehicle could easily become uncontrollable. Despite these concerns, Coolbaugh agreed to view a demonstration of the system.[13]

The recording problem was only one of the electronic issues. From the outset electronic reliability was crucial. The system would have no human maintenance once it was in orbit, and it was expected operate for up to a year (if not longer) in an incredibly harsh environment after experiencing a great deal of vibration during launch. Turning to the RCA Electronic Components Laboratory, Coolbaugh probed for the best method to ensure reliability, but there were no conclusive answers. The best that the facility could do was recommend that Coolbaugh meet with members of RCA's David Sarnoff Laboratory to discuss the matter and solicit the views of J. M. West, vice president of Bell Laboratories. Reliability, moreover, was also a function of quality control during production, and no one knew what level of control would ensure the satellite's reliability.[14]

Coolbaugh met frequently with his superiors to discuss progress and plans, but they most often talked about funding. On March 15, 1954, he met with Col. John Kay at ARDC headquarters. After outlining his plans for the satellite program, Coolbaugh briefed Kay on lab work and the problems that arose due to lack of funds. Aware of the financial problem, Kay promised his best efforts to release money from the budget for FY 1955, but he did not anticipate substantial amounts until FY 1956, so he urged the use of laboratory funds as much as possible. Coolbaugh also inquired about contracting both the Rome Air Development Center and the Air Force Cambridge Research Center to conduct research and development in communication and atmosphere work, respectively.[15]

To tap into laboratory funds, Coolbaugh exploited a loophole that his friend Pete Murray had pointed out to him in the spring of 1953. The fiscal year started on July 1; every year the labs reviewed their budgets in late March and early April to be sure they had applied all their allotted funds. If they had excess funds on July 1, their budget would fall by that amount for the next fiscal year. This situation created an incentive for the labs to help Coolbaugh in late spring and use up their excess funds. To speed up this process Coolbaugh and Capt. Buford B. Biggs, who worked in the procurement section of the WADC, developed a system that accelerated the processing of the paperwork. Normally a work order could take days to weeks to be processed. Under the Coolbaugh-Biggs system, it could be processed in an hour. The result was the most effective use possible of scarce resources.[16]

With the laboratories at the WADC working on technical issues, Coolbaugh turned to writing the development plan for the satellite system. Aware that he could expect little money for FY 1955, he assumed that lab resources could pay for technical efforts in the first year. Creative funding could be stretched only so far, however. Once he required a contractor and began actual development, the financial situation would be critical. As his planning progressed, Coolbaugh came to realize that the selection of a contractor was going to create a major problem. There were no potential candidates with any experience in this field. Normally the air force chose a prime contractor, who then selected subcontractors to develop components, with the air force approving them after a review process. Companies generally took on subcontractors with whom they had previous experience or who had submitted the lowest bid. But since no single firm could develop a satellite, the regular development process wouldn't work. After much deliberation Coolbaugh turned to the resourceful Captain Biggs.[17]

Biggs advised that they try a new approach. First, he argued that they should look not for a single contractor but for a team approach. Due to the unique technical hurdles involved, only a strong team of contractors had a chance of success, and the contract should be awarded accordingly. Second, Biggs suggested a change in how the contract bidding would be done. Following teams' submission of their initial bids, the best two or three would receive a study contract to prepare definitive proposals, which would reflect a better developed understanding of both the task involved and the strengths of the contractor-subcontractor relationship. Both Major Greene and the ARDC headquarters accepted these unorthodox suggestions.[18]

Adopting Biggs's proposal, Coolbaugh set up a rough schedule for tendering bids. Taking into account the time needed to both produce the bids and study them, he figured that the request for proposals would go out in May or June 1955, with the naming of three finalists by October or November. The ultimate selection would then take place by the beginning of FY 1956 (July 1). To cover the costs of the proposals, ARDC headquarters agreed to grant $500,000 each for the three finalists. It assured Coolbaugh that it would include $1.5 million for the proposals and $1 million for ongoing development of subsystems in the budget for FY 1956.[19] Thus the first program funding would become available only in FY 1956.

The program underwent a few changes in mid-1954. In July it established a weapon system project office—a major advance. By this time Major Riepe had reconsidered his earlier decision concerning management and took overall charge of the program office, with Coolbaugh as technical director. The two men split the responsibilities, with Riepe handling the financial wrangling and political work and Coolbaugh concentrating on the technical issues. Ten days later the program transferred, on paper at least, to ARDC headquarters, but the office and technical activities remained at the WADC. On November 27, 1954, the air force issued System Requirement No. 5, thus officially starting to develop a satellite.[20]

Despite all this progress, serious technical hurdles remained. This became evident at the end of July 1954, when Coolbaugh witnessed a test of the new video recorder system at RCA. The device was the size of a room, far outside the design parameters, and it could not handle the tape's speed; within a minute of starting, tape spilled off the reels onto the floor at the rate of three hundred inches per second. Coolbaugh wondered whether "we had better start looking around to see who else was working on video recorders."[21] A second test failure convinced him to search for an alternative.

Huckaby observed that there were two other companies working on video recorders, both in California: Bing Crosby Enterprises and Ampex. In February 1955 Huckaby accompanied Coolbaugh to California on a fact-finding tour. Bing Crosby Enterprises had nothing to offer, but the talks with Ampex were productive. To founder A. M. Poniatoff and Chief Engineer Charles Ginsburg, Coolbaugh laid out his requirements: the system had to be as small as possible, use little power, and be able to record a signal of 4.5 Mc and, hopefully another signal at 6 Mc. When told about the unsuccessful tests at RCA, Poniatoff immediately recognized the problem: RCA was moving the tape so rapidly over the recording heads that the system could not handle the speed. Ampex had the opposite approach, spinning the recording heads rapidly to "paint" the information across the tape. With the smaller and lighter heads spinning rapidly, the tape moved slowly and was under control. After watching a demonstration of the machinery, Coolbaugh signed a contract with Poniatoff to provide progress reports on the Ampex system to keep Coolbaugh's office up to date, with RAND footing the bill.[22]

Coolbaugh now pursued two methods of powering the satellite: nuclear and solar. The first was the nuclear power plant that RAND had called for. Scientists working on the nuclear-powered aircraft project were sure that a small reactor capable of producing one to five kilowatts was feasible. They put Coolbaugh in contact with Atomic International, a division of North American Aviation. This firm had been designing small reactors for domestic use and gave him a plan for such a system. Coolbaugh also began investigating solar power in September 1954. He contacted the Electronic Components Laboratory and convinced it to start experiments with photovoltaic crystals. The man in charge of crystals for the lab, Don Reynolds, persuaded Coolbaugh to support his research with cadmium sulfide crystals, which promised to produce more electrical power per square inch than conventional crystals. By the end of 1954 the first large cadmium sulfide crystals were produced. In the long term such crystals in solar arrays would solve many electrical problems.

The solar array was practical only if the satellite could store power for periods when it was out of direct sunlight. Batteries were the obvious choice. In early 1954 the best that the industry could produce was a battery that could store ten watt hours for every pound of battery. This was far too inefficient for a satellite. To get enough power storage, smaller and more efficient batteries were needed, but no research in this field was being done. So Coolbaugh contacted various manufacturers; in less than two years this produced a tenfold improvement in battery efficiency. By 1958 the ws-117l program had combined solar cell technology and improvements in battery capacity, which has became the norm for satellite power.[23]

Early in 1955 the satellite program gained access to new facilities for research and development. The Rome Air Development Center in New York State joined up to support work on the satellite's communication elements. This included all the ground-to-space microwave systems for relaying data and command instructions between the satellite and ground stations. The air force's Cambridge Research Center, located at Hanscom Air Force Base outside of Bedford, Massachusetts, also became a partner. Already active in space research as part of VANGUARD, the Cambridge Center examined the atmospheric conditions in which a satellite would travel and operate. This effort helped to determine the main threats to the satellite and thus how robust it had to be and also assisted the researchers in defining vehicle characteristics and identifying any special requirements.[24]

While components formed an ongoing focus of Coolbaugh's program, the bureaucratic system brought its own burdens. Caught within the USAF bureaucracy, Coolbaugh found that a great deal of effort was devoted to briefing his superior officers. This was necessary to keep them current on progress and, since money was a major problem in the first years of the program, to encourage financial support. Occasionally these briefings did produce good results. Coolbaugh's greatest convert to satellite reconnaissance was Gen. Donald Putt. Described as a "space cadet," a term of endearment for the enthusiastic supporters of space programs, Putt was ws-117l's highest-ranking supporter. Behind the scenes he exerted a great deal of influence on its behalf.[25] Unfortunately General Putt was the exception, not the rule.

In March 1955, one month after the TCP report appeared, Coolbaugh gave a crucial program briefing to a group that included many members of the top air force brass—most notably, Assistant Secretary of Defense Donald Quarles. Displaying models of the reactor and the inertial guidance system, he told his audience about the program, its goals, and its current status. Although Putt and two or three other people showed marked interest, Quarles was not enthusiastic. Except for a few sharp questions his lack of enthusiasm was manifest to everyone. According to Hall, Quarles did not oppose the program but saw it as a long-term effort, and so he was unwilling to give it too much support.

At roughly the same time, Coolbaugh and Riepe learned that less funding than they had expected was available. The FY 1956 budget for the program had dropped from $2.5 million to only $1.5 million, just enough to cover the design studies. Whether this shortfall resulted from Quarles's lack of interest, a desire to cut corners, or Coolbaugh's success in using excess funds from the labs, there was money only for contract bidding.[26]

On March 15, 1955, the air force formally issued General Operational Requirement No. 80, calling for the development of a strategic reconnaissance satellite system and providing technical requirements for it.[27] This General Operational Requirement represented top-level approval for the program and laid out for the first time its full and formal requirements. It defined a satellite that could survey the world's entire surface, determine a potential enemy's ability to wage war, supply information for national intelligence, and warn of an attack on the United States. The images had to be clear enough to allow interpreters to identify airfields, cities, factories, and other strategically important structures. There were also requirements for gathering intelligence (through photographs, signals intelligence, and infrared and other sensor systems) and for ground stations to monitor the satellite and process the data.[28]

Following the document's release, a staff team began preparing for contract bidding, and by the spring of 1955 this group had expanded. Riepe was still director of the satellite project; Coolbaugh remained technical director and specialized in power supply and propulsion. First Lt. John C. Herther took control of guidance and stabilization, and Capt. William O. Troetschel was placed in charge of communications and command and control. This foursome brought the program together during the initial development.[29]

In May these senior officials held a Request-for-Proposal meeting with prospective contractors at Wright Field. They invited representatives from all of the top aircraft and electronics companies and briefed attendees on the proposed program and technical work to date. Riepe also explained the contract-bidding process and the crucial role of the chosen team of contractors. No single firm could develop the program, so the final choice would depend heavily on the quality of the team of contractor and subcontractors.

The favored combination was Douglas Aircraft and Bell Telephone Laboratories because of their outstanding reputations and joint experience on military contracts, but they declined to bid. Quarles was a former vice president at Bell, and when J. M. West from the company asked him about the program Quarles strongly opposed the firm's bidding on the contract. In his opinion the air force would not be undertaking a serious satellite effort for at least a decade, so the program would do little more than keep the idea alive until then.[30]

The fact that the senior Department of Defense official in charge of research and development was not enthusiastic explains a great deal about the program's slow evolution, despite the staff's best efforts. Without the support of highly placed officials, the program was doomed to progress slowly. Although no documentary evidence has been found to show that Quarles specifically blocked funding during the first year of the program, the circumstantial case is clear. His attitude toward the program before the launch of *Sputnik* in 1957 can be described as indifferent at best.

Second Stage: Western Development Division (1955–57)

The summer and fall of 1955 was a busy time for the ws-117L team. Besides the many meetings with labs and consultants, the group had to deal with the transfer of ws-117L to the wdd and evaluation of the first-stage contract bids. News of the transfer arrived as the team prepared to evaluate the bids. As mentioned, the wdd had already taken over the ATLAS missile program. Since the ATLAS was likely to be WS-117L's first-stage booster, the transfer would facilitate consultation on matters relating to both programs. The head of the WDD, Maj. Gen. Bernard Schriever, had advocated the move in an attempt to prevent likely competition for resources.

Accordingly staff members from the WDD attended the deliberations on contractors' bids.[31] Its senior representative, Capt. Robert Truax (USN), received a warm reception. Called the "first space cadet" because of his long-standing interest in space, his personality and similar outlook impressed his counterparts. A small group of scientists accompanied him from Ramo-Woolridge, a private company that provided scientific and technical support to the WDD, much as the various laboratories did at the WADC. The Ramo-Woolridge contingent, under Dr. Robert Cornog, came across as arrogant and condescending. Overemphasizing the "superior" support that they would provide, they alienated every air force officer present. Their questions during the contract presentations were also insulting. Despite the tension between them and the WS-117L group, and the pending transfer of the program to the WDD, the contract proposals went forward. The evaluation team consisted of representatives of all the laboratories supporting the program at the WADC, as they were the most familiar with the various components. The three most successful bidders—Lockheed–CBS Laboratories, Martin Aircraft–IBM, and RCA–North American Aviation—received contracts for follow-on studies.[32]

On October 10, 1955, WS-117L was formally transferred to the WDD, effective early in 1956. Lt. Gen. Thomas Power, commander of ARDC, authorized the move, overriding the objections of virtually all the general officers of the WADC and his own staff. The need for a close link between the WS-117L and the ATLAS programs weighed heavily in his decision. The staff at the WADC was not happy about Ramo-Woolridge's taking over its program; its members worried, quite legitimately, that progress would slow once WS-117L no longer had air force supervision and Ramo-Woolridge tried to expand its expertise.[33]

The WS-117L team suspected that it would be difficult for Ramo-Woolridge and the WDD to support both a satellite and an ICBM effort and that the pairing of efforts would in fact dilute the resources available. General Schriever agreed, and Ramo-Woolridge did not enter the program. WADC personnel working with the WS-117L and the air force labs continued under Schriever's direction at the WDD.[34]

The move to WDD ended the program's first phase. Technical studies and development continued unabated, taking on greater importance as a contractor came on board in 1956. Budget and support remained problematic. By moving west the program was closer to fruition. With the backing of the WDD and Schriever's ICBM team, the WS-117L took on high priority. From October 1955 to October 1957 it moved slowly toward completion of a proto-type system.[35]

The order for the transfer of the WS-117L program to the WDD took effect in February 1956. The transition was fast and seamless largely because most of the WADC's staff, including key members such as Coolbaugh, Herther, and Troetschel, went along. Over-all control of the program was vested in Col. Otto J. Glasser, who was responsible for both WS-117L and the ATLAS program. Daily control rested with the WS-117L's new office head, Captain Truax, who arranged support from higher levels of authority.[36] The net result of this division of labor was that Truax, through daily control of the office, oversaw most things directly related to the satellite program itself.[37]

The first major task for the new WDD office was to complete the plan for developing the program. This involved settling myriad issues and conducting many briefings to win support for the program's budget.[38] The plan was to explain the system's goals, its expected completion date, technical details and specifications for equipment, and how the components would work together. Part of the plan explained the tasks that remained before the system's first flight and its subsequent operational deployment; it also authorized the steps to reach these goals. As such it was a crucial document, especially with regard to obtaining funding.

Colonel Glasser insisted on completion of all basic planning by April 1, 1956, to allow for the filing of the development plan before the next budget year started. This demand put a great deal of pressure on the WDD team to fit the program's elements into a coherent description. The result was a plan that called for a first orbit by May 1959, with complete operational capability by the third quarter of 1963. The document described the satellite and supporting equipment. Exclusive of facility costs, research and development would require roughly $114.7 million. General Schriever received the plan on April 2, 1956, approved it, and passed it to the office of General Power, who gave it his blessing about three weeks later.[39]

While writing the development plan, the WDD was also busy with contract issues. The three in-depth proposals for the satellite program were due in March 1956. The program brought in many original staff members from the WADC to help it make informed judgments. Of the three presentations, Lockheed's seemed the most intriguing. Its engineers believed that adding a second stage to the ATLAS could allow the payload to increase by 10,000 to 15,000 pounds. According to Curtis Peebles, this proposal called for two separate satellites: a pioneer version, weighing no more than about 3,500 pounds, and an advanced version of roughly 7,800 pounds. Peebles's descriptions are somewhat sketchy. With the development of the advanced satellite, operational reconnaissance would be possible. The WDD could not give Lockheed the contract immediately, however, because funding would only follow the approval of a development plan.[40]

Air Force Headquarters issued this formal approval on July 24, 1956, and a development directive was issued on August 3. But financial problems continued. The plan called for $39.7 million for fiscal year 1957, but the program received only a maximum of $3 million, less than 10 percent of what had been requested. The ARDC invoked "severe limitations" on the budget for FY 1957 as justification, but even it admitted to underfunding the satellite program.[41] The WDD did, however, receive enough money to award Lockheed the contract to develop the first American reconnaissance satellite. The delay in assigning the contract—until October 29, 1956—reflected the difficult financial situation.[42]

Lockheed, the prime contractor for WS-117L, now had the daunting task of bringing a complicated system together under Jack

Carter, head of the project, and his chief scientist, Louie Ride-nour, who had presented Lockheed's proposal to the WDD staff. Most of the early problems at Lockheed came not from the satellite itself but from the usual start-up challenges: inexperienced managers, miscommunication, and problems with suppliers. The early delivery of HUSTLER engines to Lockheed best illustrates this problem. Bell Aircraft hoped to have all engines delivered before Lockheed had made final modifications so it could bill its customer for later changes. Careful management by Carter and the Lockheed team, as well as the WDD, ironed out the problems, and the engines arrived on time.[43]

A small army of subcontractors supported this enterprise behind the scenes. Some, such as Ampex and RCA, had helped develop the initial components. The new ones included some of the biggest laboratories in the United States. For example, CBS Laboratories under Dr. Peter Goldmark took on development of the read-out system (part of the photographic subsystem), while Dr. Charles Draper's Instrumentation Laboratory at MIT worked on the satellite's guidance and stabilization equipment. These subcontractors played major roles in the program, although much of their work was behind the scenes.[44]

Funding continued to limit progress, however, and the WDD found it impossible to circumvent these restrictions. The stumbling block remained Air Force Secretary Donald Quarles, who controlled the rate of development—providing only a trickle of funds and forbidding construction of components past the design stage without his approval. The program could not prepare mock-ups of the satellite or its components, let alone actual experimental vehicles. Truax remembered being in Quarles's office when the secretary told him, "I don't mean to throw too much cold water on the program, but I don't want any tin-bending yet."[45]

Coolbaugh's insightful memoirs end in early March 1957. He moved temporarily to the THOR IRBM program, and his only contact with WS-117L as a THOR representative occurred in early 1958. Although he did play a role in CORONA and the DISCOVERER program, he did not return to WS-117L until January 1, 1960. Unfortunately Perry's "History of Satellite Reconnaissance" is a poor substitute; it tells us little about daily activities or overall progress before October 1957.[46]

It is possible, however, to piece together fragments of the remaining WS-117L story before *Sputnik*. We know, for example, that financial problems continued. General Schriever believed that he had figured out the air force's resistance during his attempts to find backing for the program in 1957. He blamed both Quarles and the administration's emphasis on "space for peace" over national security.

First, of all people to oppose the program, Quarles seemed an unlikely suspect. He had pushed for the IGY scientific satellite and produced the first draft statement on Outer Space Policy (NSC 5520). He saw satellite reconnaissance as beneficial and important in the long run but favored low-risk technologies and mistrusted projects that had not demonstrated complete reliability. VANGUARD would prove the technology and provide legal precedent, so he was willing to wait. Thus he was not against WS-117L, or even reconnaissance satellites, just hesitant to devote considerable resources at that time. This inclination was reinforced by his strong support of the administration's desire to cut defense costs. An untried program that promised results years in the future was a marginal priority; further study was a relatively inexpensive alternative that ensured some progress. His lack of enthusiastic support, both in meetings and in funding battles during the first three years, grew out of these views. As Eisenhower's representative and the man charged with research and development, his ambivalence is emblematic of the fact that the WS-117L program did not enjoy the support of the top-level military officers and civilian leadership of the air force.[47]

Second, the administration's stance of "space for peace" also helped to generate program restraint. By April 1957 Schriever was blaming this policy for the inertia in funding. The core of the problem rested with the idea of separating the military and civilian space programs and the legal question of satellite overflight. VANGUARD was created to establish the principle of freedom of space in a manner least likely to antagonize the Soviet Union. There were other reasons for separating the programs—

most notably Eisenhower's desire to prevent interference between the military and civilian programs. Hence the administration decided to keep VANGUARD exclusively civilian.

To this end both military proposals for scientific satellites and military attempts to accelerate the VANGUARD program proved pointless.[48] Both the army and the air force claimed that they could orbit a satellite sooner than the VANGUARD effort and more economically, but the administration consistently blocked them. The army was pushing its own REDSTONE-JUPITER missile combination to launch a scientific satellite. The air force also entered the effort, pushing a variation of the WS-117L program that used nonreconnaissance components to launch a scientific satellite. Unsuccessful attempts to promote this endeavor detracted from the WS-117L effort in general. The irony here is that while the administration publicly urged "space for peace," it never clearly defined *peaceful*, aiming to keep the Soviets pliable vis-à-vis freedom of space.[49]

To circumvent the funding problem Schriever developed a plan that may have inspired development of the CORONA program. Understanding that the United States had to advance the program for intelligence purposes, in April 1957 he ordered "Fritz" Oder to devise a policy to boost the status of the air force's satellite program. In the meantime he maintained pressure on the air force to supply financial support through regular channels. This effort failed, however, when he had to accept a reduction following the first review of the development plan in April 1957. Sure that the administration was intent on sacrificing the program, he reevaluated its schedule, making cuts wherever possible. By July he had placed severe spending ceilings on the Lockheed contract, which further delayed the program. It became clear that WS-117L was in serious peril.[50]

Oder's financial solution went to the heart of "space for peace." Since financial difficulties arose from the administration's reluctance to pay for an expensive endeavor that might endanger U.S.-Soviet relations should it become public knowledge, the solution was to make WS-117L vanish. Oder's scheme—with the code name "Second Story"—was a clever piece of sleight of hand that rested on three key principles: satellite reconnaissance had to be covert, the CIA had to take an active part in it, and the effort required a massive infusion of money.

The plan was simple: the air force should publicly cancel WS-117L, but the CIA would covertly reactivate it. To provide cover for the program the air force would establish a major scientific satellite program as a follow-up to VANGUARD, which would explain the WDD's satellite efforts. The CIA in turn would keep the effort secret and, using the WDD for technical purposes, provide the satellites for an active reconnaissance program. If the plan succeeded, it would never compromise "space for peace," and the administration could support it.[51] With the plan in hand, Schriever secretly approached key members of the administration in June 1957 to gain their support. Among those he briefed were General Putt, James Killian, Edwin Land, and Richard Bissell from the CIA, who went on to run the CORONA program.[52]

Despite Schriever's attempts to increase funding to $10 million, the administration refused to bend, though it extended permission to procure items that had long lead times. Schriever felt he had to approve tentatively a schedule that would put Second Story into practice. This required that General Putt "request" that the air force develop a scientific satellite program to replace VANGUARD, backing it up with a proposal for a Ballistic Missile Division for such an effort. This division would have to be in place by the start of September 1957. To facilitate the plan, and to ensure it would interest the White House, Maj. Gen. Andrew Goodpaster (the president's military aide) and several other members of the White House staff were briefed in August. In light of later developments it is likely, but not certain, that the president received a briefing.[53]

By late September Schriever's program had bogged down. The need to coordinate efforts with numerous officials in both the scientific and the military programs while maintaining secrecy proved too difficult. Even as the Stewart Committee, which had recommended VANGUARD over other military proposals, was undergoing reactivation to help plan for a follow-up program, Schriever faced signs of trouble. The secrecy on which he depended was on the verge of evaporating. A consultant with the Department of Defense, while working on a memorandum calling for a national policy for space and totally unaware of Schriever's plans to save WS-117L, stumbled on a 1956 proposal to use the WS-117L program to develop a scientific satellite. This was the original submission of January 16, 1956, proposed as an option for the IGY satellite use of all of WS-117L's nonreconnaissance elements.[54]

The consultant raised several questions over the feasibility of using a military satellite in a scientific role, and his rejection of the idea endangered Second Story. From Schriever's perspective, if the scientific program came under attack, that would compromise the cover story. With the consultant totally unaware of Second Story and its secrecy, Schriever devised a three-stage plan to move the project forward. First, the Ballistic Missile Division had to develop a detailed proposal for a scientific satellite that he could present to the Department of Defense.[55] It would have to demonstrate not only scientific value but air force unity on the matter. Second, Schriever needed a public-relations policy to manage information about the project's scientific and covert intelligence sides. Third, he had to reassure the Stewart Committee that the scientific program would work and be beneficial.[56]

Schriever's gamble to increase funding for the WS-117L soon proved futile, for within a month the Soviet Union had orbited the first satellite, demonstrating technical feasibility. The administration and the air force, under harsh criticism for the failure of American efforts, immediately increased funding for space research and development. In effect the launch of *Sputnik* on October 4, 1957, solved the air force's budget problems by removing all the justifications for slowing efforts. With the Soviet success providing strong motivation, the WS-117L program was to leap forward dramatically in October 1957.

Due to the slow declassification process, the story of WS-117L is incomplete in some key areas. The largest element missing from the record is a description of the satellite project on the eve of *Sputnik*. Still we can reconstruct what the program was aiming for in 1957. The satellite development plan at this stage differed little from the original concept, which used a television system that stored images on magnetic tape for relay to Earth. Yet it was clear by August 1957 that this configuration would be technically unfeasible. Predicting resolution of such low quality as to negate the program's value, RCA convinced Lockheed and the WS-117L team to reject this approach. The television system gave way to a camera with more conventional film, which would receive processing on board the satellite before scanning to allow transmission. Although the resolution would not equal that available from looking at the photo directly, it was better than the television would have produced.[57] As early as 1956, however, the RAND corporation had discussed a camera system that would return the film to Earth, but the WS-117L team had rejected it, owing to technical problems with recovery.[58]

The plan had been for the program to progress in stages. The original Lockheed proposal of 1956 called for two distinct phases: a test phase, or pioneering system, to validate components and hone the process of intelligence gathering, and a second phase to transition to operational capability.[59] Several historians argue that the program actually involved three phases. The initial test phase, using a THOR-AGENA launch configuration, would begin in November 1958. This system would give way in June 1959 to phase 2, in which an ATLAS-AGENA system would test higher payload launches and longer-term orbits. Finally a third phase was to start in March 1960 and consist of three satellite systems: a pioneer photographic-reconnaissance system with a six-inch focal-length camera lens, an advanced photographic system using a thirty-six-inch lens, and a long-term surveillance system; clearly the emphasis was on the photographic systems.[60] However, these satellites would contain not only photographic-reconnaissance systems but also electronic intelligence and later infrared sensors. These dates contradict the 1956 development plan, which anticipated an initial test series beginning in 1959 and achievement of operational capability in mid-1963. The difference in dates is striking, given that funding was almost nonexistent. One can conclude only that the phases I describe derive either from speculation and are totally incorrect or from accelerated post-*Sputnik* plans—probably the latter. There are indications that these dates may reflect planning after October 4, 1957. For example, Oder's account refers to cam-era designations that clearly relate to the program after *Sputnik*.[61]

Confusion over dates is the smallest problem facing histori-ans. Some accounts refer to systems that were clearly not part of the original satellite concept. Robert Divine's account describes the WS-117L system as having three satellites: DISCOVERER, a system for film recovery; SAMOS, to relay images; and MIDAS, an infrared early-warning satellite. MIDAS and SAMOS were def-initely descendants of the WS-117L system, but Divine's descrip-tions seem to assume post-*Sputnik* program data because these names emerged only after October 1957. DISCOVERER actually may have been the initial test system for WS-117L or part of Schrie-ver's Second Story project.[62] Either way it definitely was part of the program *after Sputnik*.

The satellite program, as it evolved in the years leading up to *Sputnik*, was a coherent one. It arose out of RAND studies and tried to remain true to its original design concept. The program was extremely ambitious and called for a level of technology not yet available in a reliable form. Consequently, as technical problems appeared, they necessitated new technology and/or adaptation of existing technology. In the meantime, however, the program ran into serious external problems, especially the absence of financial and political support. These factors directly encumbered develop-ment. Unlike the technical issues, these difficulties were beyond the control of people working on the program; only other exter-nal factors could change them. That didn't happen until October 1957, but when it did it would have profound impact on WS-117L.

[6]
Satellite Photography, Film Return, and the Birth of CORONA (1957–58)

The successful launching of a satellite instrument is bound to be a
spectacular event, causing worldwide sensation.
—Paul Kecskameti

The Soviet launch of *Sputnik* on October 4, 1957, had far-reaching
consequences. No other single event had exerted so much domestic pressure on the Eisenhower administration. In the wake of
the beeping satellite, the political backlash and the whole "missile gap" controversy shook the foundations of American confidence and security.[1]

A President under Pressure (October 1957)

To recount the entire debate over *Sputnik*'s implications or to
explore the various positions on the issue is not necessary for this
study since it is amply covered by historians elsewhere.[2] The controversy's basic elements, however, can help us grasp *Sputnik*'s
impact on the WS-117L program. The missile gap antagonized relations between the White House, Congress (particularly the Democrats), and the military. Just as they had done with the "bomber
gap," the military and its legislative supporters urged increased
defense funding in response to *Sputnik*. Exploiting the press and
fear of Soviet technological "superiority," the military (especially
the air force) and several elected officials tried to use the opportunity for maximum political benefit. One of the most vocal critics
of the administration was again Senator Stuart Symington, who
used information that the air force leaked to him. Charging (cor-

rectly, as the previous chapter showed) that Eisenhower's economy measures had caused the satellite program to fall behind, he argued that the United States was also lagging in ICBM development. Accusations from Symington and others, such as Senate Majority Leader Lyndon Johnson (D-TX), placed the administration on the defensive. The press, sensing a "big story," picked up the rhetoric and intensified the attacks.[3]

At the heart of these scathing critiques lay the argument that inadequate defense spending had squandered American technological superiority and thus left the country vulnerable to the Soviet Union. Republican policy in general and the president in particular were to blame for the Soviet space coup.[4] As with the debate over the bomber gap, the military went before Congress to request more money. Beginning in November 1957 Senator Johnson gave the military an open venue for its attacks through his inquiry into satellite and missile programs. During these hearings defense spending received a thorough airing. Both Gen. Nathan Twining, chairman of the Joint Chiefs of Staff, and air force chief of staff Gen. Thomas White argued that the United States was not behind in ICBM development but had to increase spending for bombers and missile development over the long term to prevent a Soviet lead. Many pundits in the media supported their claims, predicting a major discrepancy in ICBMs. The *New York Times* journalist Joseph Alsop predicted that by 1963 the Soviets would have 2,000 ICBMs—a far cry from the 130 that experts expected the United States to have by 1962. The press helped to stoke a damaging set of accusations against Eisenhower, whom it characterized as a "do-nothing" president whose lack of resolve led to this crisis. Congressional leaders, invoking dire warnings of the Soviets' being far ahead, sought more money for expanded programs.[5]

The press and Congress's virulent attacks shocked Eisenhower, who had no idea that the nation's perception of itself was so fragile. Many Americans could not grasp the idea that another country could surpass the United States in science and technology. *Sputnik* forced many of them to question their assumptions.[6] The administration's response to *Sputnik* was not very astute. Eisenhower thought the Soviet threat to be rather insignificant. Having the advantage of u-2 imagery during the bomber and missile controversy, Eisenhower knew there was no gap in favor of the Soviet Union. The forays by the u-2 revealed the locations of launching facilities and kept the Americans generally abreast of Soviet missile work. Unfortunately the president felt that he could not reveal this publicly and endanger the u-2 as an intelligence source. To reveal it to the public would humiliate and provoke the Soviet Union to take action. Thus the president refrained from refuting the doomsayers in the Senate.[7]

But this did not stop him from looking into the matter further. On October 8, 1957, shortly after the launch of *Sputnik*, he requested clarification of the status of U.S. missile and space efforts (both civilian and military). He asked both Gen. Robert Cutler, his special assistant for national security affairs, and Deputy Secretary of Defense Donald Quarles to look at three issues for him. First, he wanted information about studies of the development of guided missiles since 1953. Second, he hoped for an account of work on Earth satellites during that period: progress reports, recommendations, and program priorities. Third, always worrying about fiscal issues, he insisted on receiving "the chronology regarding the costs of the program."[8] In essence he wanted to know why the United States had not yet launched a satellite.

Having activated the government machinery to answer his questions, the chief executive turned his attention to calming the public. He and key members of his cabinet sought to create the impression of "business as usual."[9] Saying that the satellite did not raise his apprehensions by "one iota," Eisenhower maintained that *Sputnik* did not endanger U.S. security and that there was no crisis.[10] At the prompting of members of the Science Advisory Committee of the Office of Defense Management, he took the opportunity to increase public awareness about the need for science and education for American youth rather than increase the defense budget. Through speeches and meetings with leading scientists, he stressed that his administration valued technological developments and that the best minds available were advising him on these matters.

He also took several steps to blunt the attacks from the military and Democrats. First, in October 1957 he named James Killian, who had shaped the TCP report, to the new post of special assistant for science and technology and to lead the President's Science Advisory Committee. Second, in early February 1958, he asked that committee to recommend the outlines and organization of a space program. This move led eventually to the creation of the National Aeronautics and Space Administration, which incorporated existing elements, such as the old National Advisory Committee on Aeronautics, into a coherent civilian space program.[11]

Third, Eisenhower refused to accelerate the VANGUARD program in the wake of *Sputnik*. He had been blocking military attempts to do so prior to the launch of *Sputnik*, as it would have meant using military technology to implement the program. That approach ran counter to his goal of using civilian satellites to establish the legal precedent of the freedom of space. It also seemed unnecessary since the Americans were never in competition with the Soviets to be first in space. The scientific mission remained in place. Since VANGUARD had every prospect of success, Eisenhower saw no reason to advance the effort through either schedule changes or use of military boosters, as long as progress remained satisfactory. Fourth, he asked for briefings on the progress of the military satellite program in response to his questions of October 8, 1957.[12]

Photography and Film Return (March 1956–October 1957)

The historical record of WS-117L becomes increasingly unclear after *Sputnik*, partly because of the flurry of activity that followed it and partly due to the scarcity of information about developments.[13] The gaps in our understanding are ironic and unfortunate, for the most pivotal moment of the whole story occurred between October 1957 and March 1958, a period of fundamental reorganization. On October 6, 1957 WS-117L had consisted of a single satellite, but in November 1958 the Department of Defense revealed that it consisted of three satellites: DISCOVERER, SENTRY, and MIDAS. DISCOVERER was a research-and-development satellite for testing system components and for research in various areas, including biomedical studies. In reality this was the cover for another reconnaissance satellite, CORONA, which used a system that returned film to Earth for processing. The film readout system, the core of the WS-117L concept, was dubbed SENTRY, an innovation discussed in detail in chapter 7. Finally, the infrared detection system designed to warn of a Soviet missile attack, originally part of WS-117L, was given its own satellite, called MIDAS.[14]

The historians Jeffery Richelson and Robert Divine refer to these three programs as subparts of the entire effort and as being in place prior to *Sputnik*. However, they do not explain how or when a single satellite suddenly became three divergent systems. Most scholars have argued that the film-return system was part of WS-117L before *Sputnik*. They portray it as part of the overall program, a continuation of the WS-117L concept; its separate development and operation became essential after it seemed apparent that WS-117L could not meet its schedule. These theories, however, are derived from information before declassification. No historian has provided a specific account of when the program added film return or of the impact this had.

Before exploring *Sputnik*'s effects on U.S. satellite reconnaissance, we should place the origins and development of the CORONA concept in its proper context. It was the idea of a second satellite program, revolving around film return, that emerged in early 1956 and was ready for serious consideration on the eve of *Sputnik*. CORONA joined WS-117L in *Sputnik*'s wake, and its association with the earlier effort was brief. With this development in mind we can see *Sputnik*'s impact for what it was: a catalyst to American satellite reconnaissance.

RAND was not idle following release of the FEEDBACK report in 1954. Rather it expanded its staff and kept current about the latest developments in what it still considered its program. It was not until early in 1956, however, that its team became vocal again about satellite reconnaissance. Aware of problems with WS-117L's film system and of new developments in the field, the group reconsidered a satellite that returned its film to Earth for processing, even though it had rejected the prospect a few years earlier.

Richard C. Raymond, a physicist and expert in information theory and a member of RAND's Electronics Division, looked at how much information was theoretically available from intelligence photographs and negatives. When comparing WS-117L's film-readout system then in development (similar to television technology of the period) and the amount of information available from direct examination of photos and negatives, he came to a startling conclusion. Assuming that both types of images came from the same altitude, the standard film would produce roughly two orders of magnitude more data. To make matters worse, the radio bandwidth for transmitting data was very narrow, so relays to Earth would be very slow. Television imagery, then, would be far inferior to photography. Admittedly no one had expected that the television system could compete with normal photography, but no one had realized just how much information was at stake.[15]

In light of this revelation and of advances in reentry technology that came from ICBM research and development, RAND's president, Frank Collbohm, with the help of a RAND researcher named Brownlee W. Haydon, prepared a formal recommendation to the air force's air staff in March 1956. "Photographic Reconnaissance Satellites" called for development of a recoverable satellite system. RAND withdrew the recommendation, however, within a few weeks for reasons that remain unclear. According to Merton Davies, one of the firm's experts, the materials relating to the decision were destroyed after the withdrawal, so the reasons may never be known.. Raymond, however, recalled that the air force was concerned about the time lag between the taking of the photo and the information becoming available. WS-117L offered the theoretical prospect of quick access to the data, almost in real time. A film return or drop-film system, as RAND proposed, would mean a built-in delay between intelligence gathering and film recovery. Furthermore no one knew yet if recovery was feasible. Until an in-depth study comparing WS-117L and the drop-film proposal could be completed, the air force had no reason to accept this proposal. Hence later that same year (1956) it contracted with Lockheed to develop the WS-117L system.[16]

Nevertheless RAND continued to explore the technology of returning film to Earth. By June 1956 enough progress had been made that it could issue a research memorandum on the physical recovery of payloads. J. H. Huntzicker and H. A. Lieske's "Physical Recovery of Satellite Payloads—A Preliminary Investigation" clearly demonstrated that a film-returning satellite system had strong merit. First and foremost, it could collect more information than a relay satellite. Photos offered greater resolution (and therefore more information), and captured data about the environment the satellite worked in would increase scientific knowledge and provide an understanding of, among other things, the conditions affecting film, the camera, and orbits. Second, film return was likely to provide intelligence sooner. The complicated nature of the WS-117L system, and its projected 1960s operational date, meant that it would be years before it was ready to work. The simple film-recovery satellite could be utilized almost immediately for intelligence gathering. Third, the recovery technology for film was seen as a major step toward manned flight in space.[17]

The successful recovery of a capsule from orbit depended on three elements. First, the satellite trajectory or orbital path would need modification so that the satellite would reenter the atmosphere at a predetermined point and have time for successful landing and recovery. Second, it was essential to protect the payload from heat to ensure the integrity of intelligence materials. Third, a means of locating and retrieving the payload was crucial.

Huntzicker and Lieske found that current technology could meet all three conditions. Orbital adjustments were the easiest goal to achieve. Since all orbits decay over time, recovery necessitated premature orbit decay, which was a function mainly of atmospheric drag slowing the satellite, forcing it to orbit lower in the atmosphere. The use of a rocket to decelerate the satellite (or part of it) would suffice. In addition it was possible to adjust the direction of travel and thus to identify a general area of recovery.[18]

Protecting the payload from the heat of reentry presented the greatest challenge. ICBM research and development had already begun to work through this problem; warheads, like the satellite payload, had to reenter the atmosphere and survive to reach their targets. The two programs' similarities meant that one could borrow technology from the other. Heating the surface of the reentry vehicle was a function of the satellite's flight path, speed, size, and surface material and area. To minimize the heat of reentry, only a part of the satellite should return to Earth. Furthermore layers of fiberglass insulation and metal could protect the payload and diffuse the heat that builds up as the satellite descends into denser layers of the atmosphere. Since the most intense heating would probably occur in the first thirty seconds of reentry, the RAND group recommended separation of the outer heat shield only after this point. This shield would have built up a great deal of heat, and its ejection would prevent that energy from radiating to the inner layers of protection.[19]

Detecting and recovering the payload seemed straightforward in comparison. A recovery system would probably weigh roughly 227.6 pounds, including the film payload (about 50 pounds), the reentry rocket, insulation, beacon, power supply, and parachute. Since most of the weight—the heat shield and reentry rocket—would jettison shortly after reentry, recovery would be a great deal simpler. While a parachute covered final descent, increasing the time in the air, radio beacons would indicate location. RAND expected easy payload detection and recovery. Being able to predict roughly the area of recovery made the odds of success very good. The report did not define the method of recovery; rather it anticipated use of conventional ships and aircraft.[20]

RAND's work on returning film did not stop there. For RAND the inherent simplicity and value of recovered film meant giving the recovery system high priority, but the air force and the WS-117L program offices were not very supportive.[21] At this time WS-117L was not yet under way, contract selection had just started, no funding had arrived, Assistant Secretary Quarles was offering no support, and he and other senior officials in the air force had still not accepted development of the FEEDBACK satellite. A satellite that returned photos to Earth via a capsule seemed even more radical than WS-117L. Until recovery of a payload proved feasible and a detailed comparison between FEEDBACK and film return was available, the military rejected the idea.[22]

RAND, however, continued to support the idea of a film-return satellite. Two of its main thinkers on satellites had picked up the idea and kept the work alive. Merton E. Davies, a longtime "space cadet," and Amrom H. Katz, a convert to satellite reconnaissance who had years of experience in photo interpretation, quickly grasped the merit of Huntzicker and Lieske's study.[23] They continued work on the drop-film camera system throughout 1956 and early 1957, and slowly an advanced satellite took shape.

Two external groups helped them greatly. First, the Boston University Optical Research Lab provided the basic idea for a new, higher resolution camera. On February 19, 1957, several of the Boston Lab staff, including Walter Levison, Dr. Duncan Macdonald, and James Baker, briefed Davies and Katz on their ideas for creating a camera capable of providing 120-degree coverage. Second, their concept gained reinforcement with information they obtained in early March from Fred Willcox, vice president of Fairchild Camera and Instrument Corporation. Willcox had been designing a panoramic camera to provide wide-area coverage for aircraft reconnaissance. By spinning the camera he increased considerably the area that it could photograph. Davies and Katz adapted both these concepts—higher resolution and wider coverage—into a new system. The "Hyac" (for high acuity) camera promised a theoretical resolution of one hundred lines per millimeter of film (roughly ten times the resolution of a Second World War system).[24]

By early 1957 the duo had merged this high-resolution panoramic camera with a long-focal-length lens to generate a new high-altitude system. Since the main problem of photography from space was the difficulty of obtaining high-resolution shots that covered large areas, the combination seemed to solve the principal problem that Katz found in the WS-117L system. In early 1957 Davies and Katz believed that they had developed the best means of accelerating the WS-117L program by merging the Hyac camera, the film-return concept, and the THOR IRBM as a first-stage booster.[25]

To accomplish this goal, however, the WS-117L program had to overcome air force resistance, and do it with enough momentum to sustain development after the initial impetus had dissipated. *Sputnik* created the essential catalyst to release the bureaucratic brakes and affected the American effort in four ways. First, *Sputnik* demonstrated feasibility; it forced skeptics within Defense and the administration—Quarles, for example—to recognize the possibilities of satellite reconnaissance. Second, *Sputnik* had legal repercussions. The American VANGUARD scientific satellite program had aimed to establish the legal principle of freedom of space. With *Sputnik*'s success the entire issue became moot since now the Soviet Union could hardly complain about a U.S. satellite flying over its territory. Ironically, by being first in space the Soviets had done the WS-117L program a sizable favor. Third, the Soviet triumph drove home the need for intelligence. Clearly the Soviet Union was close to developing a feasible ICBM design. The possibility of a deployed ICBM force increased the risk of surprise attack and illustrated the unequivocal necessity to track Soviet activities closely for possible signs of deployment. Such tracking would be easiest during construction of launch sites, when the large amount of activity would be observable.

Fourth, *Sputnik* heated the political temperature in the United States. A firestorm quickly obscured the entire effort on reconnaissance satellites. While partisan attacks and congressional inquiries dominated the landscape, the military services again looked to turn the situation to their benefit. Having warned of the increased Soviet threat, the services (particularly the air force) saw *Sputnik* as vindication. Taking advantage of the situation, all three services sought carte blanche for large-scale missile and satellite programs, along with a bigger share of the defense budget. Congressional testimony and the press became vehicles toward these ends, and a plethora of proposals swamped and distorted the political process. Beneath this layer of infighting, the WS-117L program was doing its best to ride out the deeper currents that *Sputnik* created.

Many sought to accelerate the program, but these attempts ran into trouble. Concern about WS-117L first emerged in a report by the President's Board of Consultants on Foreign Intelligence Activities. This board was created by Executive Order 10656 on February 6, 1956, to monitor foreign intelligence activities, review them, and give recommendations to the president on how to improve intelligence.[26] The board's report on October 24, 1957, displayed serious doubts about the American ability to gain intelligence on the Soviet Union. The board insisted on a thorough review of the advanced reconnaissance systems then under development. It received briefings on two such systems—the WS-117L satellite and the Project Oxcart/SR 71 aircraft, a supersonic successor to the U-2 but designed to be more invisible to radar—and determined that the programs were not receiving adequate support. Neither seemed likely to provide intelligence soon, and Killian worried particularly about the satellite program, pleading for clear decisions on its priorities.[27]

According to secondary sources, the board's report of October 1957 concluded that the WS-117L program was far from accomplishing its goals. The board, which included Killian and Edwin Land, discovered several reasons for its tardiness. The original development timetable was far too optimistic and proved unrealizable. Air force mismanagement had resulted in a lack of focus, and shortage of funds only made the situation worse. Finally, the program was at the edge of current technology. The board found its camera/radio-relay system to be far below the necessary level for practical intelligence gathering. The camera, with its short focal length, offered only one-hundred-foot ground resolution. Integrating it with the narrow-bandwidth radio transmission would substantially slow return of the imagery to Earth. Although the timetable called for active testing by late 1959, 1960 seemed the earliest realistic date. The board also learned that the air force, already mismanaging the program, was publicizing it in an effort to win more money for its research—clearly an attempt to exploit the situation that *Sputnik* created.[28]

Air force efforts frustrated Killian and Land and their colleagues on the panel. In order to speed acquisition of vital intelligence about Soviet missile development, the group called for a simpler accelerated WS-117L program that could be operational as soon as possible. In conjunction with this recommendation, it asked for an interim intelligence system. Borrowing RAND's concept of a film-return satellite, this second satellite system seemed less complicated and therefore more practical for rapid development than WS-117L. Killian and Land argued that the CIA should develop the system with support from the air force, in much the same way as the two bodies had developed the U-2; this method would avoid the inefficiency and bungling of the air force effort.[29] Thus as the initial shock of *Sputnik* began to dissipate, a small group of enthusiasts was already proposing a second system, to return film to Earth in a capsule. This system would not replace WS-117L but merely accelerate collection of intelligence on the Soviet Union.

In late 1957, as the President's Board of Consultants on Foreign Intelligence Activities reported on its findings and questions concerning an interim program were being discussed, WS-117L supporters sought ways to overcome its funding problems. Initially they thought that the response to *Sputnik* would overcome resistance to satellite reconnaissance and free up funding, but surprisingly there was still opposition at high levels of the air force and the Department of Defense and skepticism from the White House itself. This grew out of concerns not directly related to the program but rather with respect to the very idea of an "open" reconnaissance effort. Although classified, the WS-117L project was not "black" in terms of its secrecy, as there had been newspapers articles about the program. This raised concerns of a Soviet reaction to reconnaissance satellites.[30]

Eisenhower's reaction intensified the reluctance to accelerate the program. Resenting inferences that the administration had ignored space and missile development, he would not approve any form of acceleration if it appeared to be a "crash effort." He worried lest the press see such action as an admission of culpability. The strongest opposition at Defense came, not surprisingly, from Deputy Secretary Quarles. Still opposing rapid development, he blocked the acceleration plan on October 16, 1957. To circumvent his decision, General Putt secured permission from Air Force Secretary J. H. Douglas to present his case for a speed-up directly to the new secretary of defense, Neil McElroy, who took over on October 9, 1957. This briefing, on October 29, finally overcame Quarles's opposition. On November 1 Secretary McElroy authorized the program, which was to proceed "at the maximum rate consistent with good management."[31]

The WS-117L program did receive a much-needed financial boost in the wake of *Sputnik*. The pre-*Sputnik* budget of the air force's

Ballistic Missile Division had stood at $991 million, virtually all of it for ICBM development, which was on a tight schedule. For fiscal year 1958, however, the WS-117L's budget increased from $13.6 million to $65.8 million—not enough to overcome years of neglect but a step in the right direction.[32]

Attempts to solve the funding problems were bolstered by the National Security Council when it issued NSC Action 1846, "Priorities for Ballistic Missiles and Satellite Programs," on January 22, 1958. This document helped the situation by opening the way for placing satellite reconnaissance at a higher level of importance. Now the secretary of defense could assign high priority to satellite programs that had "key political, scientific, psychological or military import."[33] On June 20, NSC 5814, "U.S. Policy on Outer Space," placed the SENTRY/SAMOS and DISCOVERER programs on the priority list. By giving satellite reconnaissance such a high designation, the president made it clear that WS-117L was vital to national interests, without creating the appearance of a crash program. This stance helped to eliminate problems relating to funding and access to resources and personnel.[34]

Neither RAND nor the air force had been idle. In the months between the initial suggestion of a film-returning satellite system in March 1956 and the launch of Sputnik, RAND kept the idea alive. By November 1957 Katz and Davies were ready to submit a more comprehensive report on its feasibility and a formal recommendation for its development, which they did on November 12. The document was the product of a massive joint effort within RAND and represented one of its most significant research memoranda.

It attacked the problems in gathering intelligence much more systematically than earlier studies had. Looking at satellite reconnaissance in terms of intelligence requirements and the current state of technology, the report urged a series of progressively more complicated systems. Calling for better acquisition in three major types of intelligence (warning of attacks, estimation of enemy capabilities, and targeting information), especially vis-à-vis Soviet ICBMs, Davies and Katz described aerial reconnaissance in terms of four distinct levels of information gathered. The difference between these levels was based on the size of the area a satellite's cameras covered, ranging from thousands of square miles to tens of square miles.[35]

The first level, A, was a system for searching a large area and covering vast expanses quickly. Poor resolution would prevent detection of small objects, but A was useful for monitoring activities and providing data that would permit precise coverage by more advanced systems. In short it was the foundation level of intelligence gathering. Level B would allow identification of many major installations, aircraft on runways, and minor lines of communication (smaller roads and shorter railway and canal lines). Level C would facilitate photography of specific objects on airfields and in industrial zones, and level D was to provide even finer photos. With a scale that was very small, this information would satisfy most of the technical intelligence needs. The trade-off, of course, was that as resolution increased, coverage decreased, so it would take more time to cover the entire Soviet Union.[36]

Davies and Katz proposed a reconnaissance system that could move from level A to level D in a continuum of technical improvements. Unlike the ws-117L satellite, it used a spin-stabilized satellite that rotated the camera shutter across the earth's surface, exposing the film with each revolution. The initial satellites would probably employ an Aerecon camera with a focal length of twelve inches to test the system and to begin photographic reconnaissance. Since the entire camera rotated with the satellite, stability was less of a concern. The camera, sweeping across the line of flight, would photograph large areas. The film had to move through the camera fast enough to compensate for the satellite's speed so as to prevent ground blurring. The shutter speed (1/4,000 second) meant that substantial changes in altitude, speed, and so on would not degrade the image. Once it proved itself reliable, the more advanced Baker twenty-four-inch camera would come into use. Recovery of the film would require the de-orbiting of the entire satellite. Katz and Davies projected that the resolution for such a system would be about sixty feet, with a single shot covering about 18,000 square miles. Thus a reel holding five hundred feet of film could photograph about 4 million square miles, or roughly half of the Soviet land mass. All major targets, including airfields, lines of communications, and urban and industrial areas, would be visible.[37]

The basic configuration of the system that Katz and Davies proposed was both simple and practical, and it harnessed existing technology, greatly reducing development time. This prototype was to serve as the foundation for more complicated satellite systems. Second- or third-generation satellites would probably have camera lenses of longer focal length (twenty-four and thirty-six inches) for higher resolution and greater film capacity. This first system, however, needed little time for development; a satellite could be in orbit within a year of the date of the contract. Davies and Katz hoped that this system would augment and support the ws-117L program by supplying intelligence in the short term while work proceeded on the follow-on system.[38]

Davies and Katz described clearly both what the intelligence the system could provide immediately and its growth potential. The initial system would provide level A, or area, surveillance with resolution comparable to air force reconnaissance aircraft. Through repeated coverage, it could monitor changes in any area that indicated construction or development and indicate patterns of activity. In terms of growth potential, improvements to camera and lens systems would enhance resolution, allowing for far superior photos and thus better data. The initial lens of twelve-inch focal length would probably give way to a thirty-six-inch system within eighteen months. Three years after the initial contract improvements might lead to ten-foot resolution.[39]

With such a study behind it, RAND's recommendation for development of a film-returning satellite was compelling. Because of the expected intelligence dividend from such a system, RAND suggested reprogramming of ws-117L to emphasize tasks requiring space-to-ground communication capability—specifically level A coverage. Because of its lower resolution and slower relay capability ws-117L would be ideal for area surveillance, identifying targets for the superior resolution of the second and third generations.[40]

The ws-117L staff and the Ballistic Missile Division paid attention to Davies and Katz's studies for RAND. In November 1957, when the Ballistic Missile Division asked it to consider modifications to speed up the program, Lockheed's Missile and Space Division proposed two alternatives. The first included use of intermediate range ballistic missiles (IRBMS) as a booster for testing and early flights. The second incorporated a film-returning satellite along the lines of RAND's proposal. The combination would mean even an earlier satellite launch.[41]

Lockheed included these ideas when it briefed the air force and General Schriever on program acceleration late in 1957. Schriever was impressed enough to request a formal proposal for hastening the program. Missile and Space Division's proposal of January 6, 1958, drew heavily on both the RAND work and some of the original concepts that Lockheed had developed for the WS-117L contract bid in 1956 but had shelved because of funding limits.[42]

Lockheed still had to meet air force requirements for visual reconnaissance as well as for electronic intelligence and infrared systems. The plan called for a series of development phases. The first phase would test a series of satellites to verify the orbital capabilities of the booster/AGENA system. The plan slated the inaugural flight in this series for October 1958, the final one for February 1960. The THOR IRBM—available in larger numbers than the Atlas ICBM and thus able to sustain more flights—was to see service in some of these tests. While gauging orbital capability, this phase would also test the visual, electronic intelligence, and infrared systems. The second phase, Program II, was to achieve pioneer visual reconnaissance; the intention was to provide resolution of at least one hundred feet and accuracy of up to one mile in mapping a location. The goal was locating airfields, major cities, and industrial centers, with the first flight likely in March 1960.[43]

The proposed RAND system received a development position as part of Program II, namely Program II-A. It was to be a 7,200-pound second stage launched from the top of a THOR IRBM. Using a RAND-style method of returning film, II-A would supply imagery equal to the early stages of WS-117L. The launch schedule mentioned six satellites between January and July 1959.[44]

Program IV of the WS-117L system was the more advanced photographic configuration, with a total resolution approaching twenty feet and location accuracy on the order of one-half mile.

Finally, Program VII featured the infrared system. Initially part of the overall ws-117L program, it had become the basis of its own network of satellites to transmit early warning of an attack. Programs III, VI, and VIII remain classified but probably made up the electronic intelligence satellite system, called ferret satellites in the vernacular of intelligence operations. Programs VI and VIII seem to have included better visual and more advanced ferret capabilities for continuous surveillance.[45]

The most crucial questions about technical feasibility concerned the new requirements of Program II-A, notably the first-stage IRBM booster and the recovery of film from space. The booster was straightforward. The THOR, a smaller missile than the ATLAS, could not support as large a payload. Its limited lift capability of about one hundred pounds to an orbit altitude of three hundred miles was its main liability. However, when combined with the AGENA second stage, then in development for the ws-117L system, it could carry a projected payload of two hundred to four hundred pounds. The THOR had thus become a viable booster at least for testing and returning film. Therefore Lockheed's Missile and Space Division rejected the three-stage booster configuration that RAND put forward in memorandum RM-2012. A simpler two-stage version seemed adequate for a useful interim reconnaissance mission until the more advanced ws-117L system became available.[46]

The recoverable film payload was another matter. The proposed recovery system would require a method of activating the process on command over Alaska for recovery in the Pacific north of Hawaii. A heat shield using the ablation technology for ICBM reentry would protect the film capsule. Impact down range would occur probably within a circular error probability of about thirty miles, allowing air and naval units to close in and retrieve the capsule fairly quickly. This system differed from the RAND proposal just in its recovery of only the nose cone with film inside. The rest of the satellite (electronics, fuel cells, camera, and so on) remained in orbit, thus reducing the necessary amount of heat shielding to roughly sixty pounds.[47]

Except for a shift in the schedule for the first flights, the accelerated Lockheed plan did not change the original WS-117L system to any appreciable degree. It relied on the THOR IRBM and RAND's film-returning satellite. Although smaller and less powerful than the ATLAS, the THOR was available in larger numbers and more reliable. This meant a much faster pace of launchings and tests and thus quicker detection and elimination of flaws in the components. The acquisition of intelligence could begin that much sooner. The interim addition of the RAND film-return system almost in its entirety did not accelerate WS-117L, but it did speed up acquisition of intelligence. The WS-117L system remained as it was; experts simply grafted the Lockheed plan onto it to make intelligence gathering faster and to provide a testing platform. The addition of Program II-A is significant, however, for understanding the division of the WS-117L in early 1958. It was a short-term supplement that by February 1958 had separated for independent development as CORONA.

Film Return, Project II-A, and the CIA's CORONA
(November 1957–February 1958)

Even as Lockheed was drawing up plans for expediting the WS-117L, forces beyond its control were eclipsing its efforts. The RAND design proposal had circulated within the air force and among senior scientific advisors in the White House in November 1957. The ideas behind RAND's drop-film system had appeared in the October 1957 evaluation of the WS-117L program by the President's Board of Consultants, and the most vocal proponents of a dramatic change in the program, Killian and Land, strongly endorsed the idea.

Eisenhower took heed of their advice and on October 28, 1957, had his executive secretary advise Secretary of Defense McElroy and CIA Director Allen Dulles that he wanted a joint briefing on the status of advanced reconnaissance systems. Because the issue was so sensitive, McElroy (speaking on behalf of Dulles and himself) proposed oral briefings with no written records. Always security conscious, the president agreed. What is clear is that during this blackout period, running from December 5, 1957, until roughly February 28, 1958, the chief executive decided that the program needed major changes.[48]

The alteration occurred in absolute secrecy. On February 6, 1958, Killian and Land met with Dulles, McElroy, and Quarles to discuss a new proposal for a film-return satellite. The following day they met again to brief Eisenhower. They proposed removal of those portions of the ws-117l program that promised the most rapid success—namely the film-return satellite (Lockheed's Program II-A)—for independent development.[49] Unlike the ws-117l program, this was not a "pie-in-the-sky" scheme; Land made it very clear that it was a "specific small project for bona fide intelligence purposes."[50]

To be effective the whole program had to be as covert as possible, hiding "under the cloak of other activities" to reduce the political ramifications of a reconnaissance satellite. There was, however, a technical and strategic reason for this as well. The resolution of the initial camera system was likely to be inferior to that of the u-2 (resolution of fifty to one hundred feet, as opposed to approximately six inches for the u-2 Baker camera). Careful precautions were essential to guard against Soviet discovery of the spy satellite; the Soviets could easily deceive and confuse American photo interpreters with dummies and camouflage. Fortunately the capsule radiated no signals, so the greatest threat to its covert nature lay in espionage or leaks during development and operation. Land and Killian called for joint air force–cia production, with the air force dominant—an idea that Eisenhower rejected. He favored putting the cia in complete and exclusive control of all the intelligence phases, with as few people as possible knowing about it.[51]

Eisenhower's distress over satellite management lay behind his insistence on cia control. The military, particularly the air force, appeared to concentrate more on lobbying for funds than on producing the reconnaissance satellite it was already developing. All three services had advocated their own brand of space program in the wake of *Sputnik*, with the army's virtually identical to the ws-117l system. However, it was the air force, which was actually developing satellite reconnaissance, that sought to translate the ws-117l program into political clout.[52] Congressional hearings discussed the program, and details about it soon appeared in the media. A mere ten days after *Sputnik* orbited, *Aviation Week* revealed the existence of a reconnaissance satellite program under the name PIED PIPER (WS-117L's moniker during contract bidding). Giving considerable detail about the program, the article noted the participation of both RAND and Lockheed, the satellite's expected configurations, and such diverse tidbits as company nicknames for the project.[53]

Eisenhower acted decisively. In February 1958 he ordered the separation out of WS-117L's Program II-A for independent development. He stressed two priorities: rapid development for operation by spring 1959 and covert work to avoid antagonizing the Soviet Union, conceal the satellite's capabilities, preserve its intelligence value, and prevent military hindrance. Recalling the CIA's effective role in developing the U-2, Eisenhower ordered that agency to take the lead in development and to work closely with select elements of the air force.[54] This decision reflected his view that the CIA should control national intelligence, as the National Security Act intended it to. If the air force had developed the program, then it would have had considerable control over strategic intelligence. Because it tended to leak information and exaggerate Soviet capabilities, Eisenhower would not trust it to dominate collection and evaluation of intelligence. By giving the CIA control, as with the U-2 program, he expressed confidence in its ability to produce results quickly, quietly, and objectively.[55] He believed that he was initiating an interim program, but it outlasted WS-117L by many years.

The decision to accelerate satellite reconnaissance by removing Program II-A from the WS-117L effort and giving it to the CIA under the name CORONA was understandable. With few details available about Soviet missile deployments, and under a great deal of pressure from all sides to increase research and development on ICBMs and space programs, Eisenhower desperately needed a system that could begin obtaining intelligence as soon as possible.[56] At the same time, he suspected that the old WS-117L program could not do the job. The film-returning satellite offered the irresistible promise of a speedy solution.

The program that Eisenhower gave the CIA had the most lasting impact of any on U.S. intelligence. Officially canceled on February 28, 1958, Program II-A was formally restarted on April 21, 1958, when Andrew Goodpaster gave the CIA the go-ahead. For the next twenty-eight months the CORONA program underwent an intensive and often painful development effort, which included many spectacular failures before it produced success on August 19, 1960, when it managed to orbit the first successful reconnaissance satellite.[57] The satellite provided the first images from space and served as the backbone of U.S. space reconnaissance for well over a decade. People responsible for its development included Richard Bissell from the CIA (who had also run the U-2 program) and his deputy, Brig. Gen. Osmond Ritland (General Schriever's vice chief of staff).[58] Having collaborated together on the U-2, Bissell and Ritland already had a good working relationship. The CIA and the air force both made major contributions to CORONA. The CIA brought experience in developing and maintaining clandestine reconnaissance systems, along with money that no one could trace back to any specific program. The air force had missile expertise and experience in research and development, along with the thousands of person hours that it had already put into the WS-117L program.

All of this would serve as the foundation for many of CORONA's basic concepts. The air force was responsible for all aspects relating to operation of the satellite system, including development, launch, physical control, communications, and recovery of the space vehicle. Besides providing security, the CIA supplied camera systems and looked after creation of special film, developing of photos, and processing of data from them. The two sides complemented each other well largely because of Bissell's approach. Believing in a small team that facilitated fast decision making and individual responsibility, he and his staff combined their talents to maximum benefit. Bissell answered directly to Allen Dulles—a marked improvement over air force management. There was no elaborate, ponderous chain of command to contend with when the program needed swift changes.[59]

The CIA employed clever sleight of hand to make the requisite parts of the WS-117L system vanish. Bissell and his staff used a variant of Second Story, the cover program for a "scientific satellite" that Schriever proposed in 1957. On February 28, 1958, the WS-117L system formally canceled Program II-A. It so informed Lockheed and made arrangements to cover costs to date. This came as a severe shock to everyone outside the loop, and the outcry from some people, such as Katz, lent credence to the cancellation order. By complaining loudly and bitterly and pushing for a rethinking, Katz helped to convince many people that the decision was real and final. The very short list of people with "need-to-know" clearance did not include him and Davies.[60]

Since some explanation was needed for the large number of THOR-based satellite shots, the CIA decided to hide CORONA in plain sight. The WS-117L plan added a new program called DISCOVERER. The agency described it as a scientific satellite program to support future manned space efforts by providing environmental and test data as well as validating the recovery concept and other aspects of space flight. It was also to help test some WS-117L components, specifically the AGENA second stage.

Thus the public DISCOVERER program gave cover for the clandestine satellite system, as did a handful of satellites that went into orbit for biomedical test shots—to generate scientific data for public consumption to hide the gathering of intelligence. This was not CORONA's sole protection, however. The fact that the public associated WS-117L with a spy satellite helped to deflect attention from CORONA. Continuing the development of WS-117L as a follow-up system (although no public announcement took place) drew the attention of interested parties. With the visible military program eclipsing it, DISCOVERER/CORONA could develop in the shadows of the program that spawned it.[61]

As a system CORONA was far simpler than WS-117L. It achieved many firsts for the CIA and the air force: the first reconnaissance satellite to photograph the earth, the first recovery of an artificial object to have orbited the earth, and the first airborne recovery of a satellite component.[62] In its simplest form the program consisted of a two-stage launcher using the THOR IRBM as a first stage and the AGENA booster as the second. The system was to establish a near-polar orbit, allowing it to photograph all key areas of the

Soviet Union. Operating for twenty-four to forty-eight hours, it employed a timer so that it would photograph only areas within the Soviet Union, thus maximizing use of the film on board. After the film's exposure it spooled up in the nosecone of the AGENA rocket. At the appointed time, on command from a ground station, the satellite disconnected the nose cone and fired a rocket to slow it down for reentry. Once in the atmosphere and through the initial heating stage, the capsule deployed a parachute, and the armed forces recovered the entire package in the air or once it had reached the surface, somewhere in the Pacific north of Hawaii.[63]

CORONA was clearly a separate program from WS-117L. Film return was not part of the WS-117L system except from the end of 1957 to early 1958, when the program added it to speed up acquisition of intelligence. It was not an integral part of WS-117L so much as a graft onto the program. CORONA was only superficially similar to the WS-117L satellite concept. With a different style of management, a different operational plan, and a decidedly different camera system, it had few similarities to WS-117L, except in a handful of components, particularly with respect to the AGENA booster. Conceptually, however, the two programs had one central connection: without the WS-117L development program CORONA would have been beyond reach in 1958–59. It was only the initial, groundbreaking discoveries of the RAND studies and air force work that made orbiting a system like CORONA possible. Though a major success in its own right, CORONA owed a great deal to the WS-117L program that preceded it.

The CORONA program transformed U.S. intelligence gathering and the nation's understanding of the cold war. But its later history will not be discussed in any greater detail here. The reasons for this are threefold. First, CORONA was separate from the rest of WS-117L and so branches off from its story. Second, CORONA is already the subject of much high-quality work by some of the best historians in the field. Interested people may consult any of the published sources that are cited here. Third, from the start this book has sought to redress the historical imbalance between WS-117L and CORONA. For our purposes the CORONA story effectively ends here.[64]

With an understanding of how CORONA came into existence, its shape, and its scope, we can now return to WS-117L and its development from the launch of *Sputnik* to its end in the 1960s. The program underwent significant changes and suffered dramatic setbacks in the wake of *Sputnik*. In the shadow of CORONA, this part of the story of satellite reconnaissance has attracted little attention and found even less understanding.

[7]
SENTRY/SAMOS, MIDAS, and the Dissolution of WS-117L (1958–60)

> As far as the satellite [*Sputnik*] itself is concerned . . . that does not raise my apprehension one iota. I see nothing at this moment, at this stage of development, that is significant in that development as far as security is concerned.
> —Dwight D. Eisenhower

> Members of the Samos organization, engaged in an enterprise tenfold larger and more costly than *Corona*, and convinced that the highly sophisticated E-6 would shortly displace the theoretically less capable *Corona* system, tended to be a bit more superior about the older program.
> —Robert Perry

With the origins of CORONA firmly established, the development of the WS-117L satellite program in the wake of *Sputnik* can now be clearly examined. Eisenhower and the USAF continued work on WS-117L from October 1957 through the end of his second term in office, albeit with some significant changes and restructuring. The story of WS-117L post-*Sputnik* is about the attempts to straighten out management issues as it is about the technology. A new management structure, complete with the creation of a new management body in the form of the Advanced Research Projects Agency (ARPA), was tried but failed. By the end of Ike's second term in January 1961, the satellite that I have traced so carefully was all but gone. In addition to the technical problems of satellites, serious management problems continued to plague the program, and in the end, with its division into separate satellites, for all intents and purposes WS-117L ceased to exist.

As we saw in chapter in chapter 6, on February 7, 1958, to help facilitate both the WS-117L program and the secret CORONA effort, Eisenhower took significant action to straighten out problems in research and development within the military. Displeased with how the USAF had handled the program, he ordered establishment of the Advanced Research Projects Agency within the Department of Defense to manage all research and development for military space projects.[1] Prior to ARPA the services had created their own projects from scratch, which led to duplication and wasted effort. With Roy Johnson (from General Electric) as head and physicist Herbert York as chief scientist, ARPA was to coordinate research and development in the military space program and eliminate waste, duplication, and unrealistic projects. The armed forces continued to propose program requirements, but ARPA evaluated them and assigned those appropriate for development to the applicable service. It ignored proposals with interplanetary goals and instead emphasized near-earth orbit systems.[2]

The military services did not like the creation of ARPA. They worried that Johnson would operate it as an independent command, ignoring their wishes and arbitrarily imposing its will. On March 27, 1958, their worst fears were confirmed when the service secretaries learned that ARPA would bypass them. Instead of being under control of the military chiefs, ARPA answered directly to the secretary of defense and had direct access to the three agencies—the Air Research and Development Command, the Army Ballistic Missile Agency, and the Naval Ordnance Test Station—active in military space research and development. In theory direct access to these bodies gave ARPA a great deal of control over them; in reality this was far from the truth. Since ARPA had no facilities or personnel to conduct research and development, it had to depend on the agencies to do the work. ARPA received funding only to test the feasibility of systems and components, which in the case of WS-117L meant roughly $186 million in fiscal year 1959. The armed forces had to pay for systems' development, production, and deployment. ARPA emphasized streamlining and the elimination of duplication and mismanagement during research and development.[3]

ARPA established control of WS-117L relatively quickly. Secretary of Defense McElroy authorized acceleration of the program on February 24, 1958, with the proviso that it take place under ARPA's control. Four days later Johnson canceled WS-117L's film-returning portion, Program II-A, supposedly because it duplicated the ATLAS-WS-117L effort. In fact, as I explained earlier, this directive was part of the cover plan for CORONA and was the creation of Richard Bissell and his CIA staff. The same order that canceled CORONA initiated the DISCOVERER program. It authorized the use of the THOR booster as part of the engineering and system tests for WS-117L and for biomedical experiments, thereby ensuring a plausible cover for CORONA. For CORONA, ARPA thus did little more than fund the overt portion of a covert program.[4]

In May 1958 the air force transferred the WS-117L program to ARPA management—a temporary formality, as ARPA Order No. 9-58 soon returned it to the air force's Ballistic Missile Division for technical development. The division had to submit a detailed development and financial plan to ARPA, which was not impressed with the progress to date or the expected form of the WS-117L satellite. Suffering from neglect and inadequate funding, the program also had become extremely complicated. It had augmented the visual reconnaissance mission with a plethora of other functions, including a subsystem to monitor the electromagnetic spectrum, an infrared alert mechanism to warn of Soviet missile attack, and a capsule-recovery system for biomedical experiments, to say nothing of other subsystems for satellite flight, including power supplies and guidance.[5]

The WS-117L program also called for thirty-two satellite test vehicles, with the last eight probably to carry at least a rudimentary reconnaissance capability over the Soviet Union by March 1960. Initial test flights were to begin in 1959 and involve about nineteen THOR-boosted vehicles, with second-phase testing using ATLAS as the primary booster in the first half of 1960. These test vehicles formed part of the WS-117L's budget for fiscal year 1959 of $186 million. Because of the reconnaissance capabilities of the last ATLAS shots in 1960, James Killian recommended that the Department of Defense seek presidential approval prior to these launches.[6]

The complex combination of systems that composed WS-117L was a major problem. By jamming in such a wide variety of sensors and systems, the engineers had created two distinct problems. First, each component had to be as compact as possible so that everything could fit into a small package, which complicated the technical design challenges. Second, the array of equipment meant greater complexity and more opportunities for problems. It also represented a psychological barrier to the program's progress and success.

In effect the air force had decided to forgo a functional but less effective satellite in the short term for a superior system that would be unavailable for many years. Any one of the systems—photo reconnaissance, infrared warning, or electronic intelligence gathering—might have emerged far earlier if it had been the program's sole effort. The air force's desire for perfection exacerbated the many problems and complexities. The "all or nothing" mentality meant that it reached none of its goals in a timely manner. ARPA, as well as the White House's Office of the Special Assistant for Science and Technology, worried about such issues. As a member of the President's Science Advisory Committee and through his contact with Herbert York and Edwin Land, Killian requested that York and ARPA review the WS-117L program in July 1958. The classified results are not yet available, but ARPA certainly looked at the complexity and development problems.[7]

Eventually ARPA decided that one large satellite could not fulfill all of the WS-117L system's missions. The more complex elements were holding back progress on those with higher priority or less challenging technical problems. ARPA considered dividing the effort into separate programs that could advance at their own pace, which would lead to faster system development. At the same time, people active in CORONA worried about public knowledge of the WS-117L system. The air force was using a press campaign to link itself with observation satellites and thus secure funding. Information leaked to trade journals and glossy pamphlets that described DISCOVERER in some detail raised warning flags among those responsible for the security of CORONA.[8]

Bissell and Colonel Oder worried that the identification of WS-117L with space-borne reconnaissance would lead the public to conclude—correctly—that DISCOVERER had links with it. The reason for canceling the drop-film camera system and restarting it under the name DISCOVERER had been to prevent just this linkage. Fearing that a backlash would lead to cancellation of CORONA, Bissell and others pushed for some means of distinguishing it from WS-117L, resulting in a compromise. According to Curtis Peebles, on October 20, 1958, Johnson ordered a change of program names. Afraid that the designation "weapon system" (WS) had aggressive implications, he ordered the air force to stop using it.[9]

To improve management, in September 1958 ARPA also ordered the division of WS-117L into three programs. DISCOVERER, still the cover for CORONA (formerly the WS-117L's Program II-A and now under CIA control), officially remained a test bed for experiments on generic problems relating to space operations, along with biomedical support missions. The film-readout system, the core of WS-117L, it renamed SENTRY. This moniker lasted until June 1959, when Eisenhower questioned having code names with military implications and had the label changed to SAMOS.[10] The infrared system, the other key element of WS-117L, separated out to form its own satellite, MIDAS (Missile Defense Alarm System).[11]

ARPA's association with WS-117L and its descendant systems did not end there. Its supervisory role often placed it in the middle between the air force and the administration over funding for programs. To prevent rampant overspending to achieve a satellite at "any cost," ARPA's leadership often had to fight with the Bureau of the Budget and the Department of Defense for the conflicting goal of sufficient funding. For example, in the budget for fiscal year 1959, citing the need to complete the program as soon as possible, the air force sought to increase its budget from $152 million to $220 million. ARPA acted as a counterweight, arguing the budget down to $185 million and then, because of its position between the program development offices and the administration, fought to gain that money for the program.[12]

The splitting of WS-117L into three systems was a pivotal move in the history of U.S. reconnaissance satellites. By allowing the various elements to progress at their own pace, the United States could develop a wide range of future satellite options because there was no single design. The most technically challenging programs no longer prevented the country from gaining the benefit of systems that could be available earlier. It also permitted DIS-COVERER/CORONA to progress with less attention and interference from outside, and, in the long term, it would be easier and faster to modify any one program.

SENTRY/SAMOS: Readout and Film Return

The SENTRY system remained basically the same as that in the WS-117L development plan of January 1958. Fortunately ARPA required regular status reports, which give us glimpses into the shape of the program in 1958–59. By the end of January 1959 SENTRY consisted of several elements. With an operational goal of providing a timely and "continuous (visual, electronic or other) coverage of the USSR and satellite nations for surveillance purposes," the program envisioned two intelligence packages for the satellite: film relay and electronic intelligence.[13]

SENTRY was to supply equipment capable of mapping, locating targets and defenses, gathering information on military and industrial strength, and monitoring electronic signals. With an intended operational life of one year or longer, it needed a high degree of reliability. The visual reconnaissance program would take photographs and store them on film. The satellite would then convert these images into an electronic signal to relay to a ground station. Initially the satellite was to use standard photograph technology with a special system to develop film on board. After developing photos the system would scan the images and relay them to Earth. Later versions envisioned electrostatic sensors and a high-resolution television system working in conjunction with a magnetic tape-recording system. At the start the resolution would probably be twenty feet but would eventually reach five feet or less. Thus the original television system that RAND's FEEDBACK report proposed in 1954–55 remained a long-term goal.[14]

SENTRY's photographic subsystem consisted of six different cameras, E-1 through E-6, which would provide sequential improvements in performance.[15] E-1 through E-3 used readout systems, and E-4 through E-6 film return. The E-1 camera was the pioneer system and would use very fine-grained film and a lens with a focal length of six inches to produce images with a hundred-foot ground resolution. Operating over the target area for only five minutes, it would return images to three U.S. ground stations, delivering about 10 percent of the intelligence to a central analysis station within the first hour after transmission, and the rest over the next eight hours. To be effective several satellites would provide virtually constant coverage. The E-2 system followed the same pattern but had an extended operational life and used a thirty-six-inch camera lens, reducing the resolution to probably twenty feet. The E-3 was an electrostatic system that relied on television cameras and videotape to record the images for relay.[16]

SENTRY's three remaining camera configurations consisted of film-return systems, despite formal cancellation of that part of the program and its transfer to CORONA. The E-4 camera was a mapping system that duplicated the army mapping satellite ARGON. When ARGON received approval on July 21, 1959, it replaced the redundant E-4 and was grafted onto CORONA. The E-5 involved a panoramic camera with a large recovery capsule. According to Dwayne Day, it had a focal length of about sixty-six inches and a likely resolution of two feet. In reality it never achieved better than a six-foot resolution and never flew as part of the SENTRY/ SAMOS program. Going into mothballs following SAMOS's cancellation, the E-5 cameras later became part of the CORONA program. With the code name LANYARD and designation KH-6 (as part of the Key Hole satellite system), it flew only a handful of missions in 1963 before a second cancellation.[17] The E-6 camera, an upgrade of E-5, was a late addition in 1960 as part of a revamping of SAMOS. It was a recovery system with a focal length of almost thirty-six inches and a ground resolution of roughly eight feet and would probably have had full-area surveillance capability with better-quality images.[18]

The January 30, 1959, SENTRY development plan also included

some details concerning the Signals Intelligence system that was to form part of SENTRY. The ferret component was likely to progress from the most basic package with only limited capability to a sophisticated surveillance system and to collect data from electronic radiation sources in the Soviet Union. The system was to provide more accurate measurements and better location techniques while monitoring signals in the range of 30,000 to 40,000 Mc/second. This would allow the United States to establish an indication of the imminence of possible hostility as well as an electronic order of battle and data useful for technical intelligence purposes. The ferret system would store data on the direction to the signal source, its frequency, and its power, using a filter and indexing system, and then relay the information to the ground via a signal link similar to the one for the photographic system. To date very little declassified information relating to this area has become available, but even this small cache is very revealing.[19]

No other major differences are evident in the rest of the satellite configuration. Still anticipating that a nuclear power supply would give way to a solar power/battery to power the satellite, the development plan also detailed some of the other systems. Along with the airframe, propulsion, guidance, and control, it discussed ground-space communication and the complex data-processing systems. Apart from outlining equipment for data retrieval and calibration, the document called for systems to maintain accurate positional data to ensure targeting, a method for initial interpretation of data to find indicators of problems or changes, and a mechanism for saving information in a manner conducive for use by standard equipment and operating agencies.[20]

Lockheed's Missile and Space Division remained the program's primary contractor. Responsible for management and systems development, it provided the central direction and monitored the many subcontractors to ensure work of an acceptable standard. Subcontractors for the SENTRY system were varied. The rocket motor, for example, was originally a Bell Aircraft engine, and the company maintained a subcontract for any modification to it. Eastman Kodak was responsible for the design and fabrication of the visual reconnaissance equipment and for conducting simulations to facilitate system development. Airborne Instruments Laboratory developed the entire ferret package.[21]

The SENTRY program envisioned a lengthy testing process. The first launch of the film-readout system was likely in April 1960, with following tests in June, August, and October through December 1960. The recovery system would probably not begin flights until January 1961, and eight flights would complete the testing cycle. There would also be wind tunnel, environmental, and component tests. The program was expensive; for fiscal year 1959 costs totaled $108,969,000, most of it from ARPA. Lockheed received the largest amount: $66,700,000. Among the increasing costs were amounts for ATLAS missiles (up from $600,000 to $14.2 million). Planning for fiscal year 1960 indicated that expenses might reach $170,500,000. ARPA was again providing the lion's share ($160 million), and the air force the rest ($10.5 million). As before, Lockheed received the most funding ($104 million), and acquisition of ATLAS missiles amounted to the next highest cost, at $34.4 million.[22]

Despite this massive investment, the SENTRY/SAMOS system was not very successful. Although it was to be CORONA's successor, severe problems prevented it from reaching its goals. It proved incapable of achieving the high resolution necessary for productive reconnaissance. Carrying about 4,500 feet of 70-mm film, the satellite had a far lower film capacity than CORONA due to the extra space needed to develop and store the film for scanning. Once developed, the film was to go to a series of storage reels and loops to await scanning and relay to Earth. The readout system was far too slow to be practical. Scanning the film required focusing a beam of light on it through a flat window. Concentrating on only a small portion of the film at a time, the light had to move back and forth across its width to scan the entire image. Behind the film a lens/receiver system collected about 75 percent of the light that came through, multiplied it, and transformed it into an electronic signal.[23]

The transmission process was tedious. For an image at 200-mile altitude, resolving objects of fourteen feet in size, the scale in the photo amounted to 1:400,000. Thus each frame of 70-mm film covered a ground area of 270 square miles. Because of the limited bandwidth (6 Mc/second) available to transmit the imagery to Earth, the satellite would take three hours to relay five minutes of exposures, far slower than the designers had expected. Making matters worse, the scanning system was reading only white, black, and one shade of gray. The more shades of gray for which the system scanned, the longer it took to process the film. Since each ground station could accept only about sixty-two frames (roughly 16,740 square miles of target area) each day because of the short contact with the satellite, the readout system seemed far from acceptable. By comparison, one early CORONA satellite could scan about 1.5 million square miles in a day with higher resolution because the image did not degrade during transmission.[24]

The image-readout system that was the hallmark of SAMOS had a very short operational life. The first system, the E-1 (one-hundred-foot resolution), consisted of three satellites, of which the program launched only two. The first attempt failed because of a launch accident on October 11, 1960. The second (SAMOS 2) made orbit on January 31, 1961, but operated for only one month. The third never orbited. The follow-on system, consisting of the improved E-2 camera and an F-2 ferret system, was no more successful. The program built only two of these satellites, with resolution of roughly twenty feet; there followed one unsuccessful launch and the abandonment of the second satellite.

Because of the poor image quality from the SAMOS 2 satellite and a damning study by the Directorate of Defense Research and Engineering, SAMOS was canceled in 1961. The directorate's study was extremely hard on SAMOS because of weaknesses in its readout system and the concept of giving operational control of both the satellite and interpretation of intelligence to the air force. Its author, Dr. B. H. Billings, echoing the views of Herbert York and Undersecretary of the Air Force Joseph Charyk, argued for a national intelligence capability. How much influence the Billings report had on Eisenhower's decisions of 1960–61 to create the National Reconnaissance Program is not clear. What is certain, however, is that a real-time satellite had to wait until the technology could catch up to the concept.[25]

In the meantime the film-return versions of SAMOS were not much more successful. The E-5 system commenced operation in 1961. Using solar panels and a more sophisticated camera system than CORONA, it was likely to produce higher-resolution images of smaller areas. The first E-5 launch in November 1961 failed to reach orbit, and a second attempt in December orbited but with no film recovery. The third E-5 system, which went up in March 1962, also failed to provide any imagery. The program was canceled and its remaining cameras put in mothballs. The more advanced E-6 system began to fly in April 1962. Like its predecessors, it experienced a series of setbacks before finally recovering film in November 1962. The imagery was no better than the CORONA system, and the program canceled the last remnants of the SAMOS system.[26]

MIDAS: The Road to Early Warning

The MIDAS satellite was more successful than SAMOS. The November 5, 1958, ARPA Order No. 38-59 separated out MIDAS and gave it to the air force's Air Research and Development Command, with the goal of providing an operational satellite system by April 30, 1960. This order mandated several tasks, including production of two completely functional infrared satellites, continued studies of infrared tracking with the goal of creating more advanced tracking capabilities, and sustaining an experimentation program to provide physical data on infrared-spectrum phenomena to improve the engineering of the MIDAS system. To facilitate these efforts an initial $750,000 became available for the period November 1, 1958 to January 31, 1959, to cover immediate costs. For fiscal year 1959 ARPA had budgeted $22.8 million for the program; this went up for the next fiscal year to $46.9 million. This was a far cry from its treatment of SENTRY/SAMOS.[27] The division of responsibility between the ARDC and ARPA was very straightforward. While the ARDC would supply technical direction for the program, ARPA provided policy and technical guidance. Thus the ARDC had to conform to ARPA's policy direction. As with the SAMOS system, regular monthly reporting would inform ARPA of developments and needs within the program.[28]

Many officials in the Eisenhower administration certainly recognized the system's potential worth. The report by the Ad Hoc Technical Advisory Board to ARPA evaluated the MIDAS proposal in February 1959. Strongly supporting it, partially because of its value in defending the nation and its technical simplicity, the report argued for more research and advocated pursuit of the program. The Early Warning Panel, part of the President's Science Advisory Committee, also kept abreast of MIDAS and recognized its long-term value to national defense. In its March 1959 review of possible methods for early warning of a missile attack on the United States, the panel portrayed MIDAS as the most promising of the numerous options available. Although more costly than the U-2 and the Ballistic Missile Early Warning System, MIDAS was likely to have a greater range than any other proposed system because of the altitude at which it operated. The review was realistic, however, in assessing the remaining questions. Expressing concern over the effects of cloud cover and atmospheric absorption, the ability to discriminate between missiles and high-altitude jet aircraft, system reliability, and so on, the panel nevertheless felt that MIDAS deserved development. With operational costs ranging between $200 million and $600 million, it recommended only initial-phase testing until answers became available for many of the outstanding scientific research questions.[29]

The MIDAS program was as ambitious as the SENTRY/SAMOS program. By December 1959 Lockheed's Missile and Space Division expected that a prototype operational satellite would be ready in March 1962. The plan called for six satellites to be in orbit by the end of June 1962 and for this number to increase until deployment of a fully operational system (eight satellites) by December 1962. The system concept called for a web of MIDAS satellites to cover Soviet territory continuously; it would detect any launched missile at an altitude of thirty-five thousand feet. The information would go to ground stations in Britain or the North Pacific. From there it would travel to a MIDAS operational center for assessment and relay to North American Aerospace Defense Command and the White House. The satellite would identify the approximate location of a missile launch, thus allowing the United States to locate Soviet ICBM fields, the approximate direction the ICBM was taking, and the number launched.[30] The MIDAS network of eight satellites in two groups of four, orbiting at an altitude of two thousand miles, with each group moving in a different orbital plane, permitted maximum coverage with the fewest possible satellites.[31]

MIDAS used the same basic AGENA booster as the SENTRY and DISCOVERER systems but carried a different payload. The seven-unit ground system for handling data was also more complex. Three tracking and acquisition stations, all within the continental United States, would acquire precise orbital information, adjust orbits, relay commands, and monitor equipment.[32] They had links to the Technical Operations Control Center, which would monitor the three tracking stations, provide orbital data to people processing information, and schedule and control changes in satellite operations. The final three units all related to monitoring of missile launches. Two readout stations, one in the United Kingdom and one in Alaska, would monitor the satellites. They would receive warnings of Soviet missile launches and relay them to the MIDAS Operations Center, which would combine them with orbit data and accurately assess the warnings. The MIDAS Operations Center would then alert the president and the military, while also directing and coordinating the entire operational system.[33]

A Plague of Problems and the President's Panel (1957–60)

Progress on satellite reconnaissance was problematic from 1957 to 1960, with two types of problems hampering it. The first problem was technical and stemmed from the need to invent the technology and the science simultaneously. All three satellites (SAMOS, MIDAS, and CORONA) presented daunting challenges that called for technology that was years ahead of its time. We can see the problems clearly in CORONA, a far simpler system than WS-117L. From the first failed launch (*Discoverer 0*) on January 21, 1959, through to the launch attempt of *Discoverer XII* on June 29, 1960, problems plagued the CORONA program. Only the second launch, on April 13, 1959, could be considered a partial success. Once the satellite achieved orbit, ground control mistakenly activated the recovery sequence too early, and the capsule landed on the island of Spitzbergen, north of mainland Norway, where the Soviets recovered it, although they never acknowledged having done so. Since the satellite contained only instruments and telemetry equipment, the loss of the capsule did not endanger the program. The next two launches, *Discoverer XIII* and *XIV* (August 10 and 18, 1960, respectively), were successful; *XIV* returned the first photographic reconnaissance of the Soviet Union from space on August 19, 1960. System unreliability and technical problems continued for years.[34] MIDAS and SAMOS both suffered from mechanical failures as well.[35]

The second major problem facing the programs had to do still with management and funding. ARPA's control of the military space program had not improved management but created more confusion and instability, further slowing progress. Brigadier General Ritland and Lieutenant General Schriever, both intimately connected with the spy satellite programs, strongly criticized the management. ARPA's lack of leadership and direction often resulted in "inadequate or untimely decision, or the lack of a decision at all."[36] Financial difficulties made the whole situation even worse; the program received money month by month, which made long-term planning and development more difficult. In September 1959 Schriever complained bitterly to air force chief of staff Gen. Thomas D. White about this hurdle to both SAMOS and MIDAS. Because of constantly changing support levels, he wrote, "we are forced into a day-to-day type planning for conduct of the program." Consequently "establishment of a logical plan and execution of that plan in terms of procurement, scheduling, production, test and operation of SAMOS has been rendered essentially impossible."[37]

ARPA's management had proved so inept that Secretary of Defense Neil McElroy finally acted. On September 18, 1959, he ordered the removal of SENTRY/SAMOS from ARPA's control and returned it to the air force, pending acceptable development and operational plans. As soon as it learned of this decision, the Strategic Air Command indicated that it expected to gain operational control of the program upon its completion, and its leadership immediately began to prepare plans to accelerate SAMOS's development. Pushing for a broad-front approach to research and development, before the satellite's designers had finished their work SAC was generating plans for ground-support structures and training of staff. Without any conclusive evidence that the system would work, the air force had planned how to spend its extra money on the program. In short, it was "putting the cart before the horse."[38]

Management for the program did not return to the air staff directly. SAC's concept for program development and operation ran afoul of the Directorate of Defense Research and Engineering, which it had set up in 1958, with Dr. Herbert York in charge. The directorate was above ARPA in the chain of command and was responsible for authorizing military and Department of Defense research projects. With ARPA removed from program management, the Directorate of Defense Research and Engineering assumed responsibility for development. York's views on the matter led to strong clashes with SAC and the air force. He and his staff refocused the program away from image relay—the foundation of the WS-117L program since its inception in 1954—and advocated film recovery instead. Supporters of the real-time potential of film readout, especially SAC, and those favoring use of as many approaches as possible to provide intelligence as early as possible (the Ballistic Missile Division and the ARDC) fought this change. Realizing that the air force had to accept the restructuring in order to regain control of the program, the ARDC finally bowed to the directorate's pressure. The secretary of defense officially transferred SAMOS from ARPA back to the air force on November 17, 1959. The ARDC again furnished development plans and requested funds for the programs on February 18, 1960. For SAMOS the air force requested $160 million for FY 1960 and $199.9 million for FY 1961 in order to maintain at least minimum levels of research and development. To achieve the same for MIDAS, it asked for $31.1 million and $40 million for FYS 1960 and 1961, respectively, above and beyond money remaining from research and development in this area under SAMOS.[39]

The unilateral decision to change the SAMOS program to a film-return system shows that the directorate's leadership style was far more divisive than necessary. The order to restructure the program met with acquiescence, but the Ballistic Missile Division still fought to maintain the readout system. Arguing that abandoning work on it would delay a reconnaissance system until mid-1963 unless there was a rapid increase in funding, the Ballistic Missile Division stubbornly maintained SAMOS's readout capability as part of the program. Such resistance led to longer delays in program development. The infighting over funding and priorities between the readout systems (E-1 to E-3) and the film-return systems (E-4 to E-6) prevented decisive leadership in satellite development.

SAC's attempt to usurp responsibility for SAMOS before it completed research and development caused greater confusion and diluted efforts to create a single, unified program. York and his directorate saw SAC's efforts to speed up SAMOS as unrealistic and harmful in the long run. In December 1959 he openly attacked these attempts, claiming that they "inevitably interfere with the research and development program." He found SAMOS "confused and slowed down" by the concentration on an operational system in advance of establishing capability.[40] Favoring film return himself, York discouraged development of ground-support systems until there was certainty about the satellite's actual form. The Ballistic Missile Division's approach of relying on a readout satellite system and its development simultaneously with all the ground support elements did not change in the face of this opposition. It resolutely included these concepts in every presentation on its development plan.[41]

The loss of a U-2 over the Soviet Union in May 1960 reinforced the Eisenhower administration's belief in the importance of satellite reconnaissance. In January 1960 the president had asked for a presentation on the status of the reconnaissance satellite effort where the problems with SAMOS seemed self-evident. On February 5, at Land's request, York briefed Eisenhower on the progress of DISCOVERER, SAMOS, and MIDAS. Dr. George B. Kistiakowsky (the president's science advisor) and Land questioned the air force's emphasis on the SAMOS readout system. Accepting its long-term value, Land worried that it was not yet technically feasible and was rather an ultimate objective. He felt that recoverable systems should receive priority because they were likely to be available sooner and generate immediate intelligence products.[42]

A May 1960 meeting between Eisenhower, Kistiakowsky, National Security Advisor Gordon Gray, and General Goodpaster, the president's staff secretary, emphasized the management failures of the SAMOS program and its inability to meet intelligence needs. Kistiakowsky argued that its bad management and poor guidance had led to a program and a satellite that was beyond the practical feasibility of existing technology. Eisenhower questioned how such a program could spin out of control and decided that he wanted sound advice on the satellite effort.

The President's Panel (May–August 1960)

Facing problems that seemed to be delaying SAMOS and MIDAS indefinitely and a lack of progress in the DISCOVERER/CORONA program, the president took direct action to fix the problem again. The presentation on satellite reconnaissance became a review of the overall program. In May 1960 Eisenhower directed Kistiakowsky and Secretary of Defense Thomas S. Gates Jr. to form a panel to assess the SAMOS program.[43] Consisting of Joseph V. Charyk, Herbert F. York, two staff members of the President's Science Advisory Committee, and Edwin Land's Intelligence Panel, this group was to examine SAMOS's likely intelligence requirements and determine their validity. The study would also look at the technical feasibility of the planned satellite systems "in relation to requirements, development schedules and technical direction of the program, together with the effectiveness of control over the scope and characteristics of the operational systems."[44] The panel would also recommend ways to improve organization and management.

This review played an important role in the future of satellite reconnaissance in the United States. Eisenhower received the results of the study in an off-the-record meeting on August 25, 1960, and it confirmed previous findings that the SAMOS program was facing major problems.[45] Citing the U.S. Intelligence Board's statement "Intelligence Requirements for Satellite Reconnaissance" of July 5, 1960, the report concluded that the technology under development could not meet the intelligence community's requirements. The board had listed three desiderata. First, it wanted sufficient resolution in images to permit recognition of objects no greater than twenty feet on a side. This meant that the satellites needed a photographic resolution of about five to eight feet. Second, the satellite had to be able to traverse the Soviet Union fully about once every month. This would provide a general search capability, while higher-resolution systems would follow up and do repeat coverage of key locations. The follow-up system, probably with the ability to recognize five-foot objects, was to supply descriptive information on items from earlier images. Third, the board called for a system with better than five-foot resolution, able to reveal the technical characteristics of objects that it photographed. The president's panel found these requirements reasonable in light of the technology then available. Prior to the Intelligence Board's statement, the only guideline for SAMOS development had been the pronouncements of a 1958 ad hoc committee on satellite reconnaissance. However, these requirements were constantly being modified, making effective management very difficult.[46]

SAMOS was a concept far exceeding existing technology. Two different readout camera systems were under development, but their resolution was likely to be very poor. Capable of resolving objects measuring only 250 and 50 feet on a side, respectively, these cameras were far below the Intelligence Board's stipulations, and the slow relay of images to Earth compounded the problem. Studies predicted that the higher-resolution camera would take about five hundred days to cover the Soviet Union and convey the information to Earth, assuming good weather. To make matters worse, SAMOS was not likely to be ready before 1962.[47] Thus at every level the SAMOS program fell short of intelligence requirements. It had long-term promise, but the air force was ignoring more promising technology. The president's panel therefore recommended converting SAMOS to a long-term research program until the technology could catch up to the concept.[48]

The panel felt that film recovery was the only workable solution. Believing that it held great promise to deliver both very high-resolution imagery and faster coverage of the Soviet Union earlier than SAMOS, it urged reorientation of the program toward this technology. SAMOS was already developing a film-return system

(E-5 and E-6). Defying orders (when CORONA began), the air force had refused to delete them from their development plans. The panel's report noted that SAMOS was neglecting film return in favor of the readout satellite, which could not supply appropriate intelligence. It argued that with effort the recovery system could very quickly produce useful results. Its emphasis on film recovery is interesting, as there is no indication that its members knew about CORONA. They did know about DISCOVERER's problems with payload recovery, even indicating that those difficulties might be soluble. However, in light of their recommendations for SAMOS, it is unlikely that they were aware that DISCOVERER was a film-recovery reconnaissance program.[49]

The main problem was getting the air force to reorient and maintain its focus on such a system, considering its ineffective management to date. The air force had divided control between the ARDC and the Air Material Command, which caused many of the difficulties. To rectify this situation the panel's report provided three solutions, all involving centralized management. The first was the concentration of responsibility and authority within the ARDC, eliminating internal division. The second was creation of a new air force command, which would manage all research and development but would not procure systems, thus eliminating the Air Material Command from the command chain. The third option would have the president remove SAMOS from the air force entirely, placing it within another agency. The report placed greater confidence on the second or third options. Regardless of choice, however, centralization was necessary for early operational capability.[50]

According to Perry, one of the air force's chronic problems was its incessant pursuit of publicity. The panel found that the air force had failed to maximize its efforts on the satellite program because it used the project to gain funding, thereby creating widespread organizational problems. It milked the satellite program for its publicity value but actually stalled work on it to keep research and development funds flowing. For example, on September 23, 1959, the Department of Defense's Office of Public Information announced the transfer of MIDAS and SAMOS to the air force from ARPA. It even described SAMOS as a reconnaissance satellite and MIDAS as an early-warning system against ballistic missiles. The air force's drive to gain public support for its role in space overshadowed the secrecy of satellite reconnaissance. Though it was never a "black" program, release of such information could antagonize the Soviets, thereby endangering SAMOS. Ironically such leaks did help to further mask CORONA by drawing Soviet attention away from the small "scientific" elements of DISCOVERER. Moreover the air force was not complying with the secretary's explicit instructions on SAMOS. According to Col. W. G. King, project office chief for SAMOS, the air force was "deliberately obstructionist."[51]

The program's setbacks and its pursuit of technology decades ahead of the times meant that SAMOS could not fly until at least 1963. The panel's three recommendations concerning management reflected its desire to see a simplified command structure. Eisenhower readily accepted its reasoning and on October 31, 1960, removed all the commands that were causing confusion. The new line of responsibility ran from the general officer in operational control of the whole program up to the secretary of the air force, through a new secretariat-level bureau—the Office of Missile and Satellite Systems, under Air Force Undersecretary Joseph Charyk—that handled administration and liaison. This change removed the program from direct military control, thus rendering it more accountable to the administration and making satellite reconnaissance part of a national intelligence program rather than an air force effort. This shift in turn served the needs of national security, not the desires or demands of a particular service. As for the technology, Eisenhower made it perfectly clear that the film readout system—cornerstone of satellite reconnaissance since 1954—was to receive very low priority but remain a research project under very tight fiscal control.[52] He decided to emphasize film recovery, which promised better resolution and faster development; more advanced systems would emerge as follow-on satellites.[53]

On the same day (August 25, 1960) that the review panel briefed Eisenhower, just before that meeting, the president received a copy of the roll of film from *Discoverer XIV*, containing the first recon-naissance photos from space. The panel's report of that day led to the creation of the Office of Missile and Satellite Systems, forerun-ner of the National Reconnaissance Office, which the secretary of the air force ran and whose existence remained a secret until the end of the cold war. This organization took over control of opera-tions and development of reconnaissance satellites and, while not assessing intelligence, provided the materials to the consumers—mainly the CIA and the military—for assessment.[54]

The post-*Sputnik* period saw major changes to the satellite recon-naissance effort. The original system, WS-117L, experienced two major modifications. First, in late November 1957 it took on a film-return system, which later became a separate unit, CORONA, in February 1958. The need for intelligence informed the decision to run the film-return elements as a distinct satellite effort. Since WS-117L was having problems, the desire to expedite this portion of the program was logical and beneficial in the long term. The resulting CORONA program ran for years and supplied a wealth of information. It was only the first reorganization for the program.

Second, the introduction of ARPA into the management system to eliminate duplication and waste in military research was not as beneficial as some had hoped. It generated a backlash within the air force, which saw ARPA as a threat to its independent pro-grams of research and development. ARPA acted in effect as a mid-dle man between the military and the administration. Because it had its own goals and desires, its interference in development processes and its inability to manage research projects effectively meant that it just added an additional layer of red tape.

Ironically removal of SENTRY/SAMOS from ARPA only created more problems for satellite reconnaissance. The air force, free from ARPA's control, saw return of the program as vindication and a sign to move full-speed ahead on development and deployment. Jumping ahead to system deployment and integration into the various commands, the air force forgot that it had to develop the satellite first. Instead of concentrating on making an effective sys-tem, it diverted its energies into creation of accouterments, such as crew training and the creation of the infrastructure to support it. The result was a "cart before the horse" approach that created more confusion than benefit.

Division of the WS-117L system by the end of 1959 into three parts—DISCOVERER, SENTRY (later SAMOS), and MIDAS—effectively ended the original program. Despite the sound logic underlying the decision, the results were not particularly effective. DISCOVERER (really CORONA), which the CIA ran, had a simpler development process and went on to dramatic success. SENTRY, which the air force's development system swallowed whole, became so complicated that its development program lost cohesion. Instead of putting its effort and limited resources into a single satellite reconnaissance effort, the air force worked on six systems plus a huge logistical and support program for them. Thus its efforts to produce a satellite were problematic at best. Seeking more funding, the air force failed to keep its collective and corporate eye on the ball. MIDAS, which has received little discussion in the literature, fell almost to the wayside. Designed to warn of attack, it made little progress until the 1960s. One of the few accounts that discuss the results of MIDAS indicates that there were some serious problems with the program. For example, despite the frequencies that it used, it picked up the sun's heat reflecting from clouds, creating false alarms. Work continued until it evolved into the Defense Support Program's early-warning satellites, which are now in geostationary orbit of roughly 22,300 miles, using twelve-foot-long infrared telescopes to monitor the world for missile launches.[55] It is unfortunate that a great deal of this story is not yet available.

In the end the most telling information concerning the development of satellite reconnaissance remains two simple facts. First, the program that began in 1954 languished because of mismanagement. Focusing too closely on other goals and padding the budget, the air force failed to produce the satellite that the United States required to monitor the Soviets. Seeking a "perfect" satellite, capable of a wide variety of missions, the air force closed its eyes to the mission of acquiring photographic intelligence on the

Soviet Union as soon as possible. Aiming too high, the program failed to live up to its goals.

Second, the WS-117L satellite (later SENTRY and then SAMOS) was a concept years ahead of the technology. The slow relay of images, problems with resolution, and the challenges of orbiting a satellite doomed it, and by 1962 it was in mothballs. Although some of its cameras (E-6s) came out of storage for CORONA, it was, in the end, a bust. Ironically CORONA, the "short-term effort" to provide reconnaissance until SAMOS was operational, went on to great success. It was able to do this because it harnessed technology already available in a far simpler way. In addition its management concentrated on accomplishing its aim—orbiting a satellite as soon as possible—and its descendants were still in operation in 1972. The long-term implications of these programs, however, deserve serious attention.

Epilogue

WS-117L in Perspective

By the time Eisenhower left the White House, on January 20, 1961, his administration had laid the foundation for U.S. satellite reconnaissance and the space effort for the next forty years. Although not successful in and of itself, the ws-117L program played a pivotal role, and its descendants still orbit the earth today. These satellites have supplied valuable intelligence, allowing the U.S. government to chart a more informed course through the turbulence of the cold war to the present day.

How should we assess the ws-117L program? Was it a success or a failure? The design that was its hallmark from conception—the readout satellite—had only one unit orbiting before cancellation. Does this make the program a failure? The answer is no. The ws-117L system had a major impact on U.S. national security, the course of the cold war, and the future of space exploration.

Ironically it was the Soviets who pushed the right button—*Sputnik*—to accelerate a program to spy on them. In one instant the Soviets demonstrated that they had rocket capability superior to that of the Americans, that satellites were feasible, and that the Soviets needed monitoring by the United States. The Soviet propaganda effort helped to create the "missile gap" controversy. By demonstrating what appeared to be superior scientific technology, the Soviets shook American confidence. The United States had seen itself as scientifically superior and now was suddenly second best. *Sputnik* galvanized Americans into action and legitimized the notion of the free use of space by establishing the legal precedent that paved the way for U.S. satellite reconnaissance.

Unintentionally the Soviet Union had done the United States a big favor; the end result was acceleration of the WS-117L program.

As the first U.S. satellite program WS-117L broke many theoretical barriers. Its multistage booster configuration and corresponding work relating to orbital mechanics, communications, and alien environments set the stage for manned space flights. Equally groundbreaking were its research on possible satellite payloads, its requirements for weather reconnaissance, and its efforts on communication and navigation, which propelled RAND beyond direct scientific uses and created applications that are now the norm.

WS-117L's work in space and reconnaissance technology was exceedingly ambitious. Not only was it planning to orbit a satellite—before *Sputnik* the realm of science fiction and dreamers—it also expected to deliver high-resolution photography. The technical challenges were enormous, as the idea of a real-time camera system was years ahead of the technology. The complex system had to perform perfectly in a totally alien environment. Mechanical reliability was therefore deficient, as the necessary technology was still in its infancy.

Despite all the seemingly insurmountable problems WS-117L laid the foundation of every other satellite reconnaissance system in the U.S. arsenal to date. In many respects the APOLLO lunar missions were descendants of RAND's efforts. CORONA, though not part of the initial plans for the program, was the most visible sign of success. From August 1960 to May 1972 CORONA satellites were in orbit during every major foreign policy crisis, feeding vital data to the U.S. government. That CORONA could orbit successfully so soon after *Sputnik* is testimony to the WS-117L efforts.

Similar to CORONA the MIDAS satellite profoundly influenced military space activities. The goal of the MIDAS program was to produce an early-warning satellite. While not completely successful, MIDAS spawned generations of such satellites, with the Defense Support Program satellite currently orbiting and guarding against missile attacks on the United States. The SAMOS readout system also has descendants orbiting the earth.

The story of the WS-117L satellite also reveals a great deal about Eisenhower's character and administrative style. His goal as president was to preserve the nation and leave it stronger than when he took office. For the good of the nation Eisenhower embraced the u-2 and the spy satellite, and it was his vision that initiated the program. The ws-117l program was of fundamental significance, for it had the potential to help stabilize the cold war.

The original concept of ws-117l was thus achieved, albeit decades after its inception. corona was the product of the failure of the air force to develop ws-117l. Only by understanding ws-117l can we put corona into the proper context. An offshoot of the original program, it stood on the shoulders of its big brother to succeed.

The world has greatly changed since *Sputnik* went into orbit, making electronic noises. The satellite forever altered the tranquil U.S. skies. As Americans look upward at the stars at night, they can feel secure in the knowledge that reconnaissance satellites are guarding national security and keeping them safe. The saga of satellite reconnaissance and space exploration continues to evolve, and the next chapter promises to be even better.

APPENDIX

Historiography of Eisenhower and Space Reconnaissance

While most readers probably do not require an in-depth analysis of the historiography relating to space and satellite reconnaissance, some grasp of the problems in the literature can help explain why we know so little about WS-117L. The overall lack of information, and the fragmentary state of what does exist, directly affect our understanding of the birth of U.S. space-based reconnaissance. The history of the subject is difficult to research, in part because it bridges so many historical fields; it does not fit snugly into one area of American history. Rather it involves presidential history and the history of the cold war, U.S. intelligence gathering, and U.S. space science and technology. It is also hard to dig into because of the often confusing differences of opinion among authors.

The core of the problems for researchers rests within three main areas. First, the original spy satellite program falls between the cracks. While most people would expect that the CIA had a major role in its inception, the truth is quite the opposite. Up until the Soviet success with *Sputnik*, the development of space-based reconnaissance rested squarely with the U.S. Air Force. It was only in February 1958 that the CIA began to take a role, with the creation of the CIA-run CORONA satellite program, an offshoot of WS-117L. The fact that its satellite effort was so successful has also overshadowed the earlier WS-117L program, which failed to produce a working satellite. Second, increased levels of security and secrecy effectively blanketed satellite reconnaissance, preventing the "loss" of information for national security reasons. Third, those few intrepid authors working in the field prior to

1995 had to rely more on speculation and rumor than hard fact. The result was a confusing account of spy satellites' development and characteristics.

The best place to begin an examination of the historical literature on American satellite reconnaissance is with Dwight D. Eisenhower. President during WS-117L's inception and development, he was the most important figure in determining early cold war foreign policy and strategic posture. How have historians treated his presidency? As both general and president Eisenhower led the United States through some crucial transitions: the Second World War, decolonization of European empires, the rise of communism, and the emergence of the national-security state.[1]

The historiography on his administration boasts two separate and clearly defined schools of thought. The "orthodox" approach emerged during Eisenhower's years in the White House and held sway into the 1970s. This school viewed the president in a decidedly unfavorable light, portraying him as an aging hero—inarticulate, unintelligent, bland, and lacking the motivation and political skills to have a major impact on Washington and the world at large. It depicted him as little more than a puppet to John Foster Dulles, his secretary of state, or to various corporate executives and other forces within his administration. These historians admit that the public loved Eisenhower yet argue that he contributed only a sense of security and stability.[2] As Charles C. Alexander argues in *Holding the Line: The Eisenhower Era* (1975), his two terms were uneventful, dull, and boring. The vibrancy of the Kennedy administration (1961–63) reinforced this interpretation. Measured against his young and active successor, he appeared to fall well short of the ideal chief executive. Historians came to see him as merely a custodian, presiding blandly over a transitional phase between two stronger and more significant periods.[3]

By the early 1980s this interpretation was the subject of serious criticism by a revisionist school. From its humble start in the 1960s, the revisionist challenge began to dominate perceptions by the late 1970s.[4] Since then this school of thought has only consolidated its gains and extended its influence.

In contrast to the orthodox school, revisionists have had access to the monumental document collection at the Eisenhower Presidential Library in Abilene, Kansas. Members of Eisenhower's White House staff were avid diarists and note takers. By the 1970s the declassification process had made large portions of the documentary record accessible to scholars. This included the minutes of all of Eisenhower's National Security Council meetings, his personal diaries, and the Ann Whitman file, which contains a record of Oval Office business and related materials. The records of many of the president's aides augment this hoard. Together they provide the clearest picture of the daily workings of the administration of any American president.[5] With this abundance of documentary evidence, revisionist historians have produced better accounts of the administration and portray the chief executive as the exact opposite of the orthodox interpretation: decisive, intelligent, and perceptive.

There are many examples of revisionist thought, but several works stand out as especially significant. Arthur Larson, *Eisenhower: The President Nobody Knew* (1968), Herbert S. Parmet, *Eisenhower and the American Crusades* (1972), Peter Lyon, *Eisenhower: Portrait of the Hero* (1974), and Douglas Kinnard, *President Eisenhower and Strategy Management: A Study in Defense Politics* (1977) are all admirable early revisionist studies. These authors reject the old notion of a "do-nothing" president; instead they see him as an active and skilled man with a strong leadership style, who sacrificed political opportunities for self-aggrandizement to preserve the dignity of the office. His strong abilities in foreign and strategic policy particularly impressed them.[6]

The idea of an active leader in control of his administration found masterful development by Fred Greenstein in his watershed essay, "Eisenhower as an Activist President: A Look at New Evidence" (1979–80). Greenstein depicts a "hidden-handed" president who played a major role in his administration, putting his "personal stamp on public policy" through application of "a carefully thought-out conception of leadership to the conduct of his presidency." The difference between him and other presidents rested with the fact that he was not overt about his control. His style of governance, Greenstein argues, involved cultivating the appearance of being above politics and decision making.[7]

Eisenhower's leadership methods reflected his views on command, encouraging debate among his subordinates, gathering of information, and consideration of different viewpoints before he made a final decision. Once he decided on a course of action, he delegated responsibilities to individuals he trusted and respected. Preferring informal influence to open confrontation, he was willing to work via intermediaries to achieve his ends. Dulles and Sherman Adams, the White House chief of staff, are perfect examples. They served as intermediaries between the president and the public (or parts of it). This allowed Eisenhower to control events while seeming to remain aloof from them. It also deflected blame for unpopular decisions. Finally, he emphasized results over publicity. The presidency was not a political game to him but rather an office responsible to the American people. He sought to represent and serve them, with a dignity that transcended the sort of egotism that so often accompanies great power.[8]

Greenstein is not alone in identifying Eisenhower as a savvy and skilled leader. Many historians who examine his policies find him highly competent. The best examples of the revisionist school, however, remain the works of Stephen Ambrose and Craig Allen. Ambrose has become synonymous with Eisenhower studies. In books such as *Ike's Spies* (1981) and his monumental two-volume biography, *Eisenhower: Soldier, General of the Army, President-Elect, 1890–1952* (1983) and *Eisenhower: The President* (1983–84), Ambrose depicts a chief executive in control of his administration and ardently dedicated to peace. He convincingly elaborates and supports the view of a supremely competent keeper of the peace. Ambrose portrays Eisenhower as thoughtful, decisive, and pragmatic, while experimenting with unconventional methods—such as the CIA, the U-2, and covert operations—in order to avoid war while protecting the nation's interests. He details Eisenhower's deep concern about the impact of continuous heavy defense spending on the United States. Fearing both for civil liberties and economic stability, Eisenhower sought to decrease world tensions in order to preserve American domestic institutions. Ambrose's biography epitomizes the revisionist school in its meticulous attention to detail concerning Eisenhower's part in the creation of policy and the daily affairs of the White House.[9]

Craig Allen's *Eisenhower and the Mass Media: Peace, Prosperity and Prime-Time TV* (1993) suggests something of the diversity among revisionists. Examining Eisenhower's use of the media to convey his message to the American people, Allen portrays an astute political manipulator who managed the press with consummate skill in order to garner public support. Political communication was vital for Eisenhower, and he learned to use the media effectively to get his message out. This interpretation confirms the revisionist view of a chief executive acutely aware of events around him and extraordinarily adept at channeling the flow of events in ways that advanced his own agenda.[10]

The great diversity of revisionist methods and materials has resulted in a much better understanding of a dynamic leader, but the complexity of the story seems to preclude interpretive closure. Nonetheless the revisionist perspective has gained wide acceptance.[11] Eisenhower the active president certainly sets up the historical context for any study of the programs and policies that emerged during his tenure.

The main limitation remains Eisenhower's strong tendency to maintain secrecy at all costs. Even when use of such secrets could have been politically advantageous, he refused to reveal information. He carefully edited his own memoirs to hide programs that he felt should remain secret. Thus WS-117L is glaringly absent from the historical record. While it is clear that he hid the original satellite reconnaissance program from public scrutiny, his overriding concerns and his struggles certainly give us an understanding of the political context in which it appeared and developed. He worried deeply about the welfare of his nation and its long-term security. To understand WS-117L we must explore other avenues of inquiry.

It is somewhat ironic that the one agency most people associate with spy satellites and intelligence gathering, the CIA, had so small a role in the original satellite program. It joined only in the wake of *Sputnik*, and most histories of the CIA at best glance over the WS-117L program. Many authors do not even recognize that it existed. For example, Ray S. Cline's pioneering *Secrets, Spies and Scholars: Blueprint of the Essential CIA* (1976) reveals nothing about the program. Other treatments, such as Victor Marchetti and John D. Marks, *The CIA and the Cult of Intelligence* (1974), provide only a cursory discussion of satellite reconnaissance. These texts offer no detailed analysis and little sense of the program's size or importance. In fact they imply that the CORONA program was the original American effort in satellite reconnaissance. WS-117L and the air force work prior to 1958 have practically vanished from the historical record.[12]

In recent years works on the CIA have provided more information on satellite reconnaissance within the framework of intelligence needs. While giving some indication of the field's origins, these studies do not provide details concerning the program from 1946 to 1958. Rather they focus on the operational program, CORONA, and the CIA's role in the use of intelligence. The unavailability of declassified information, however, has meant that the accounts in these books are sketchy at best. John Ranelagh's *The Agency: The Rise and Decline of the CIA* (1987) is typical. It discusses the WS-117L program and the decision to give control of satellite systems to the CIA. Unfortunately it presents the entire WS-117L system as a CIA program from the start. It considers neither the program's development before 1958 nor its details or differences from other satellite efforts.[13]

More recent works on the CIA and the U.S. intelligence community also fall short. This is certainly the case with Christopher Andrew's *For the President's Eyes Only* (1995) and Rhodri Jeffreys-Jones and Christopher Andrew's *Eternal Vigilance? 50 Years of the CIA* (1997). *For the President's Eyes Only* provides a general history of the CIA, tracing American intelligence gathering since the presidency of George Washington. While appreciating Eisenhower's pivotal role in developing photographic intelligence, Andrew does not discuss satellite reconnaissance in any meaningful way. He focuses instead on covert operations and other aspects of the CIA's role in the cold war. Nor does *Eternal Vigilance* provide any significant information about Eisenhower's satellite efforts. Concentrating on the post-1960 period, this collection refers to the development of satellite reconnaissance only in passing and as it relates to the career of Richard Bissell. Although it notes the importance of satellites as a means of gathering intelligence, it offers no history of their development. Where did CORONA come from? The book offers no clues. As with earlier works, moreover, there is no mention of pre-1958 developments.[14] Jeffrey T. Richelson's *The U.S. Intelligence Community* (1999) is equally disappointing on the history of satellite reconnaissance. It considers the current state of satellite technology but provides only a two-page synopsis of developments between 1946 and 1958.[15] These histories of the agency make no mention of WS-117L prior to 1958.

The scarcity of information on WS-117L is attributable partly to intense secrecy in relation to intelligence matters and the cold war. Authorities kept top-secret military satellite projects away from public view out of concern for the Soviet response to satellite surveillance of their territory. To flaunt their development seemed an unnecessary provocation, especially once the program started to produce intelligence. If the capabilities of American photographic satellites become common knowledge, the Soviets would be able to counter their effectiveness. Concealing the program was critical to preserving its usefulness.[16] Thus, starting with the CORONA program in 1958, public discussion of the subject was virtually nonexistent, and succeeding presidents have blanketed these programs with ever greater levels of secrecy. Unfortunately the air force leaked or made publicly available some information before CORONA and the security blackout, but the Kennedy administration and its successors were very good at limiting subsequent knowledge of satellites. Thus any account emerged more from hint and rumor than from hard fact.

Works on satellite reconnaissance before declassification began to appear in the 1970s and continued sporadically until 1990. Using rumors, leaked information, and memoirs, authors tended to write impressionistic books that were confusing and lacking in both detail and supporting documentation. A mixture of popular historical works on the space program and more specialized studies, this literature does provide at least two common reference points: recognition that there was a military satellite program aiming to provide photographic intelligence on the Soviet Union and that it began during the Eisenhower administration. Most accounts mention specific elements of the "program"—the role of RAND, for example, and some specific project names: SENTRY, MIDAS, and DISCOVERER—but there is no consensus on details of programs. Ironically these projects had no connection with WS-117L until after *Sputnik*, when the space program underwent fundamental reorientation; the pre-*Sputnik* efforts effectively remained secret. Such accounts only add to the confusion about the first satellite program. Perhaps the most influential of the early pre-declassification offerings are Philip J. Klass, *Secret Sentries in Space* (1971), and Anthony Kenden, "U.S. Reconnaissance Satellite Programmes" (1978). These studies introduce the problems and ask the questions that would preoccupy a subsequent generation of researchers. Following in their wake, a number of authors, including Curtis Peebles, Robert Divine, and William Burrows, have expanded on the story to the extent that the information would allow.[17]

Of all the people writing prior to declassification, the most important is Walter A. McDougall. His monumental work, . . . *The Heavens and the Earth: A Political History of the Space Age* (1985), was the first successful account of both the military and the civilian space programs within the political context of the post-*Sputnik* space race. While admitting the difficulties that security restrictions imposed, McDougall ambitiously attempts to synthesize the divergent materials available. His discussions of cold war nuclear strategic planning and U.S. policy concerning spending on defense and research before Eisenhower provides a solid historical foundation for an understanding of the space race and thus of the reconnaissance satellite program.[18] The greatest portion of his study, however, focuses on the politics of space exploration, research, and the civilian space effort. The satellite reconnaissance program is not central to McDougall's account.

The difficulties in penetrating secrecy all began to change in 1995. On February 22 President Clinton issued Executive Order 12951, releasing satellite imagery associated with CORONA, ARGON, and LANYARD.[19] This executive order, coupled with the Freedom of Information Act (which allows for the review and declassification of documents on request), has begun to yield a more complete picture of CORONA, which the CIA began in 1958. Kevin Ruffner's CORONA: *America's First Satellite Program* (1995) combines documents and historical commentary from the CIA's history staff and the Center for the Study of Intelligence. It consists almost entirely of declassified materials for the May 1995 conference "Piercing the Curtain: CORONA and the Revolution in Intelligence" and provides a wealth of technical information about the CORONA program. Further, it includes material from the Committee on Overhead Reconnaissance, correspondence among many of the key players in the development of satellite technology, important memoranda relating to various aspects of the program, and intelligence reports based on satellite reconnaissance. Kenneth E. Greer's long-classified article, "CORONA," which had appeared only in the CIA's restricted journal, *Studies in Intelligence* (1973), makes this work particularly valuable. Unfortunately, while the information on CORONA is illuminating at many levels and is helping to provide a clear account of that satellite effort, WS-117L still remains a virtual mystery.

The confusing picture of the WS-117L satellite program began to change only with requests under the Freedom of Information Act and the presidential declassification order of 1995. The act permitted declassification of some important materials, including most of the official histories on satellite reconnaissance that Robert Perry compiled in the early 1960s, the complete Technological Capabilities Panel (Killian) Report (1955), and several of the most highly classified RAND studies on satellites, such as Project FEEDBACK. This slow process has begun to uncover new aspects of the program and supporting contextual information. Yet the available material is still far from complete. Recent trends in security classification also indicate that it will be harder and harder to gain access to this material. Although what is open to access has yielded many answers, it has also provoked a host of new questions.

Authors working since declassification obviously have had a

marked advantage. Declassified documents released in conjunction with articles relating to the American space effort have greatly increased our knowledge of the civilian program. The two-volume compilation by John M. Logsdon, *Exploring the Unknown: Selected Documents in the History of the U.S. Civil Space Program* (1995, 1996), is an example of the impact of released documents on the historiography. Articles such as R. Cargill Hall, "Origins of U.S. Space Policy: Eisenhower, Open Skies, and Freedom of Space," and Dwayne Day, "Invitation to Struggle: The History of Civilian-Military Relationships in Space," make these volumes very useful. For example, Day traces the struggle between the air force and the army over control of satellites, which caused both services to lose focus on space matters until *Sputnik* forced change. Using numerous documents, including RAND studies and various National Security Council papers, such as NSC 5520, "Draft Statement of Policy on U.S. Scientific Satellite Program" (May 20, 1955), these works mark a major step forward. They place the history of space and satellites within a wider context that helps explain the reconnaissance program.[20]

The *Exploring the Unknown* anthology and Ruffner's compilation on the CORONA project are the most significant additions to the early history of the satellite program to date and confirm the importance of the RAND studies. They do not, however, go into great detail concerning WS-117L or the reasons for its failure. Like the work on satellite reconnaissance before declassification, they reveal a few pieces of the puzzle but do not provide an overall picture of what the program entailed or of the forces affecting its development and failure. Their greatest contribution remains the documents they make available to the reader.

Some recent authors have expanded our horizons. Three books— Curtis Peebles, *The Corona Project: America's First Spy Satellites* (1997); Dwayne A. Day, John M. Logsdon, and Brian Latell, *Eye in the Sky: The Story of the Corona Spy Satellites* (1998); and Philip Taubman, *Secret Empire: Eisenhower, the CIA, and the Hidden Story of America's Space Espionage* (2003)—all examine satellite reconnaissance during the Eisenhower years. They present the most coherent account of the development of satellite reconnaissance to date. Taking into account the role of RAND and various government-sponsored studies on intelligence, they concentrate on CORONA, the operational CIA satellite system, not on the WS-117L program.[21]

Discussing the period 1945–57 only briefly, these volumes concentrate on the operational CORONA program. For both Peebles and Day the twin problems of lack of funds and technical issues slowed WS-117L development. RAND's calls for a recoverable satellite system, *Sputnik*'s success, and the need for program acceleration forced the development of CORONA. By 1961 CORONA's success and WS-117L's problems resulted in cancellation of the latter. Taubman's *Secret Empire* summarizes much of the historiography relating to WS-117L and CORONA. Focusing on Eisenhower and attempts to gain intelligence on the Soviet Union from 1946 to 1957, the writer concentrates on the U-2 and CORONA programs and mentions WS-117L in passing, mainly as background for the founding of CORONA in 1957–58. *Secret Empire* nonetheless places these programs within the Eisenhower administration.[22]

As is clearly evident, recent treatments of the subject fail to illuminate the air force's WS-117L program. They offer some surface information, such as Lockheed's role as primary contractor, but nothing of the story's particulars—personalities, problems, internal politics. Because many scholars emphasize the successful CORONA satellite effort WS-117L remains an enigma.

The relative obscurity of WS-117L, even in the historiography that one would expect to take heed of this program, creates a number of problems. Most important, while we know a great deal about Eisenhower's concerns for the safety of the nation, we know very little about how satellite reconnaissance factored into his efforts to address these issues. His need for intelligence appears in the literature of both the CIA and satellite technology, but the development of a spy satellite system seems to exist in a vacuum. Why did Eisenhower support a satellite reconnaissance program before *Sputnik*? Was it solely because he wanted photographs of the Soviet Union (as the CORONA historians would lead us to believe), or did broader national interests drive his decision? The literature does not make the link between overhead space reconnaissance and Eisenhower's underlying concerns about economic health as the foundation of national security.

The pre-CORONA foundations of satellite reconnaissance—namely the WS-117L program—receives almost no attention. WS-117L is largely uncharted territory. Several historians look at efforts by RAND and the Defense Department to spur development of satellite intelligence, but this effort is no more than a starting point. Thus this book goes beyond that frame of reference into terra incognita. It uses the WS-117L program as a vehicle to show that Eisenhower was a visionary and activist president who saw spy satellites as more than a means to solve a difficult problem in intelligence collection. He thought of them as a way to generate knowledge that would facilitate his management of the cold war. This volume fills the void that surrounds the WS-117L program. It establishes the parameters of the air force's satellite effort, thus filling a large gap in the literature on the cold war, Eisenhower, and satellite reconnaissance.

NOTES

In an effort to provide both proper citation and a consistent flow of information for the reader, this book uses the Kate Turabian style for citation. Unfortunately, for some document collections—specifically the materials at the Eisenhower Library—there is no universally accepted method of citation. Due to the vast quantity of materials that I use from that source, and the length of full citations for some of the collections, I provide the following list of short forms. Unless I indicate otherwise, all special collections documents are from the Dwight D. Eisenhower Presidential Library, Abilene, Kansas.

AWF, Admin. Series: Dwight D. Eisenhower, Papers as President of the United States, 1953–61 (Ann Whitman File), Administration Series

AWF, A. W. Diary Series: Dwight D. Eisenhower, Papers as President of the United States, 1953–61 (Ann Whitman File), Ann Whitman Diary Series

AWF, DDE Diary Series: Dwight D. Eisenhower, Papers as President of the United States, 1953–61 (Ann Whitman File), DDE [Eisenhower] Diary Series

AWF, NSC Series: Dwight D. Eisenhower, Papers as President of the United States, 1953–61 (Ann Whitman File), NSC Series

DDE Personal Diary: Dwight D. Eisenhower, Diaries, 1935–38, 1942, 1948–53, 1966, 1968, 1969

Hagerty Diary: James C. Hagerty, Press Secretary to the President, Papers, 1953–61, Diary Entries

Harlow Records: Bryce N. Harlow, Records, 1953–61

Hazlett Papers: Edward E. "Swede" Hazlett, Papers, 1941–65

PSAC Records: U.S. President's Science Advisory Committee, Records, 1957–61

Quarles Papers: Donald A. Quarles, Papers, 1952–59

RG 59, GRDS: RG 59, General Records of the Department of State, National Archives and Records Administration II, College Park, Maryland

RG 340, ROSAF: RG 340, Records of the Office of the Secretary of the Air Force, General Correspondence—Secret, Confidential, National Archives and Records Administration II, College Park, Maryland

SPI, GWU: Space Policy Institute, George Washington University, Washington DC

SSB/GCSWS: Spy Satellites Box (copies; originals in the National Security Archive), Gregg Centre for the Study of War and Society, University of New Brunswick, Canada

WHO, NSCSP, Exec. Sect. Subject File Series: White House Office, National Security Council Staff, Papers, 1948–61, Executive Secretary's Subject File Series

WHO, OSANSA, NSC Series, Briefing Notes Subseries: White House Office, Office of the Special Assistant for National Security Affairs, Records, 1952–61, NSC Series, Briefing Notes Subseries

WHO, OSANSA, NSC Series, Policy Papers Subseries: White House Office, Office of the Special Assistant for National Security Affairs, Records, 1952–61, NSC Series, Policy Papers Subseries

WHO, OSANSA, NSC Series, Status of Projects Subseries: White House Office, Office of the Special Assistant for National Security Affairs, Records, 1952–61, NSC Series, Status of Projects Subseries

WHO, OSANSA, NSC Series, Subject Subseries: White House Office, Office of the Special Assistant for National Security Affairs, Records, 1952–61, NSC Series, Subject Subseries

WHO, OSANSA, Spec. Assist. Series, Chronological Subseries: White House Office, Office of the Special Assistant for National Security Affairs, Records, 1952–61, Special Assistant Series, Chronological Subseries

WHO, OSAST: White House Office, Office of the Special Assistant for Science and Technology, Records, 1957–61

WHO, OSAST (Killian/Kistiakowsky): White House Office, Office of the Special Assistant for Science and Technology (James R. Killian and George B. Kistiakowsky), Records, 1957–61

WHO, OSS, Minnich Series: White House Office, Office of the Staff Secretary, Records, 1952–61, L. Arthur Minnich Series

WHO, OSS, SUB, ALPHA: White House Office, Office of the Staff Secretary, Records, 1952–61, Subject Series, Alphabetical Subseries

WHO, OSS, SUB, DoD Subseries: White House Office, Office of the Staff Secretary, Records, 1952–61, Subject Series, Department of Defense Subseries

1. On the Cold War

"President Eisenhower Press Conference on the U-2 Incident and Summit Conference," in Schlesinger, *Dynamics*, 2:625.

Louis Ridenour, "There Is No Defense," in Freedman, *Evolution*, 33.

1. The literature on the debate over the reasons for the Pearl Harbor failure is massive. Some of the best works on the matter are Wohlstetter, *Pearl Harbor*; Prange, *At Dawn We Slept*; Clausen and Lee, *Pearl Harbor*. For a discussion of the impact it had on the United States, see Day et al., *Eye in the Sky*, 2–3; Kaplan, *Wizards*, 9–10, 16–39; Bissell et al., *Reflections*, 92.

2. The various investigations are Roberts Commission (December 22, 1941–January 23, 1942); the Hart Inquiry (February 22, 1944–June 15, 1944); the Army Pearl Harbor Board (July 20, 1944–October 1944); the Navy Court of Inquiry (July 24, 1944–October 19, 1944); Hewitt Investigation (May 15, 1945–July 11, 1945); the Clausen Investigation (November 23, 1944–September 12, 1945); the Clarke Investigation (September 14–16, 1944, July 13–August 4, 1945); Joint Congressional Committee Investigation (November 15, 1945–July 15, 1946). For the most thor-

ough historian's assessment of the Pearl Harbor attack, see Prange, *At Dawn We Slept*, 818–21, 841–42.

3. Kaplan, *Wizards*, 33.

4. Freedman, *Evolution*, 25–30; York, *Arms*, 5–8.

5. DDE Personal Diary, January 27, 1949, box 1, file [December 13, 1948–March 5, 1951], 1; Aronsen, "Seeing Red."

6. Freedman, *Evolution*, 33–35.

7. Starting in 1945 Brodie wrote some of the most important works in the field and is considered by many to be the key thinker in this area. See Brodie, *The Atomic Bomb and American Security*; Brodie, *Absolute Weapon*; Brodie, *Strategy in the Missile Age*; Brodie, *War and Politics*; Brodie and Brodie, *From Crossbow to H-Bomb*.

8. Kaplan, *Wizards*, 24–30. See also Brodie, *War and Politics*; Brodie, *Absolute Weapon*.

9. On the oss and its dismantling, see the documents in U.S. Department of State, *Foreign Relations of the United States* [hereafter FRUS] *1945–1950*, 1–2, 15–19, 20–23, 44–46, 89–94, 108–11; Andrew, *For the President's Eyes Only*, 156–61.

10. Presidential Directive on Coordination of Foreign Intelligence Activities, FRUS: *Emergence*, 178–79.

11. For a discussion of the CIG and the creation of the CIA, see the documents in FRUS: *Emergence*, 19–21, 166–69, 316–17, 329–31, 333–34, 364–65, 518–19, 525–33, 586; Andrew, *For the President's Eyes Only*, 168–71; Gaddis, *United States*, 306–7; Leffler, *Preponderance*, 150, 175.

12. "National Intelligence Authority Directive No. 2," February 8, 1946. See also "Central Intelligence Group Administrative Order No. 3," April 19, 1946; "National Intelligence Authority Directive No. 5," July 8, 1946, all in FRUS: *Emergence*, 332, 343–44, 391–92.

13. "Memorandum from the Chief of the Interdepartmental Coordinating and Planning Staff, Central Intelligence Group (Edgar) to the Assistant Director for Reports and Estimates (Huddle)," January 13, 1947; "Memorandum From the Chief of the Intelligence Staff, Central Intelligence Group (Montague to the Assistant Director for Reports and Estimates (Huddle)," January 29, 1947, all in FRUS: *Emergence*, 458–87.

14. "Memorandum by the Director of Central Intelligence (Souers), and Enclosure," April 29, 1946; "Memorandum From the Director of Central Intelligence (Souers) to the National Intelligence Authority," June 7, 1946; "National Intelligence Authority Directive No. 7," January 2, 1947, all in FRUS: *Emergence*, 345–47, 358–63, 478–79. Leffler, *Preponderance*, 12. Freedman, *U.S. Intelligence*, 1.

15. Central Intelligence Agency, "Threats to the Security of the United States," ORE 60-48, September 28, 1948, pp. 1–11; "CIA Research Reports: The USSR 1946–1976," Harriet Irving Library (hereafter HIL), University of New Brunswick, microfilm, reel 1, frame 0387.

16. Central Intelligence Agency, "Soviet Capabilities for the Development and Production of Certain Types of Weapons and Equipment"; "CIA Research

Reports: The USSR 1946–1976," ORE 3/1, October 31, 1946, HIL, microfilm, reel 1, frame 0017.

17. Shulsky, *Silent Warfare*, 75–78.

18. "Soviet Capabilities," frame 0017.

19. Aronsen, "Seeing Red," 112, 115. "Soviet Capabilities," frame 0017–0018.

20. "Threats to the Security of the United States," pp. 1–11.

21. Ziegler and Jacobson, *Spying*, 16, 23–24.

22. Ziegler and Jacobson, *Spying*, 21–33.

23. Freedman, *Evolution*, 25–29; Kaplan, *Wizards*, 33–39; Day et al., *Eye in the Sky*, 2–4; York, *Arms*, 6–8.

24. Rear Adm. R. H. Hillenkoetter, director of Central Intelligence, memorandum for the president, "Estimate of the Status of the Russian Atomic Energy Project," July 6, 1948, CIA Research Reports: USSR 1946–1976, HIL, microfilm, reel 1, frame 0250.

25. "Threats to the Security of the United States," pp. 1–11.

26. Kaplan, *Wizards*, 39. For further information, see Burrows, *This New Ocean*, 135–36; Ziegler and Jacobson, *Spying*, 203.

27. "Estimate of the Effects of the Soviet Possession of the Atomic Bomb upon the Security of the United States and upon the Probabilities of Direct Soviet Military Action," ORE 91-49, April 6, 1950, CIA Research Reports, The Soviet Union 1946–1976, HIL, microfilm, reel II, frame 0059, pp. 3–5.

28. "The Effect of the Soviet Possession of Atomic Bombs on the Security of the United States," ORE 32–50, June 9, 1950, CIA Research Reports, The Soviet Union 1946–1976, HIL, microfilm, reel II, frame 0120, p. 2.

29. "The Effect of the Soviet Possession," pp. 2–3.

30. Koch, *Selected Estimates*, 165–73, 189–92.

31. "Basic National Security Policy, NSC 5440, Section A: Estimate of the Situation," December 14, 1954, WHO, OSANSA, NSC Series, Policy Papers Subseries, box 14, file NSC 5440/1—Basic National Security Policy; "NSC 5501" (undated), WHO, OSANSA, NSC Series, Policy Papers Subseries, box 2, file NSC 112/1—Disarmament (3); Koch, *Selected Estimates*, 165–73, 189–92.

32. Leffler, *Preponderance*, 9–10, 81–89, 96, 246–51, 331–33, 361, 364–69; LaFeber, *America, Russia, and the Cold War*, 85–94, 101–7.

33. "Memorandum for the Record," October 3, 1956, WHO, OSS, SUB, ALPHA, box 14, file Intelligence Matters (2); "148th Meeting of NSC," June 4, 1953, AWF, NSC Series, box 4, file 148th Meeting of NSC, June 4, 1953. For an excellent discussion of Soviet espionage in the United States, see Sibley, "Soviet Industrial Espionage."

34. "Discussion at the 148th Meeting of the National Security Council," June 4, 1953, AWF, NSC Series, box 4, file 148th Meeting of NSC, June 4, 1953; Aid, "The National Security Agency," 29–31; Albats, *KGB*, 37–50, 55–58; R. Cargill Hall, "Origins of U.S. Space Policy: Eisenhower, Open Skies, and Freedom of Space," in Logsdon et al., *Exploring the Unknown*, 1:215–19.

35. Aid, "The National Security Agency," 29–31; Aid and Wiebes, "Introduction."

36. Growing out of the Sky Hook balloon research program of the 1940s, projects like Moby Dick, GOPHER, and GENETRIX (WS-119L) are all similar in nature. They all called for the use of balloons to carry cameras and other equipment either for scientific research of the upper atmosphere or in support of intelligence operations against the Soviet Union. The culmination of these efforts was the 1956 GENETRIX program, a clearly defined photographic reconnaissance effort. "Memorandum of Conference with the President," December 28, 1955, WHO, OSS, SUB, ALPHA, box 14, file Intelligence Matters (1); Gaddis, "Learning to Live with Transparency: The Emergence of a Reconnaissance Satellite Regime," in *The Long Peace*, 197. For a well-documented account, see Peebles, *Moby Dick Project*, 1–4, 28–99.

37. For a detailed discussion of the importance of the NSA and SIGINT, see Aid, "The National Security Agency"; Aid and Wiebes, "Introduction"; Bamford, *Body of Secrets*.

38. Richelson, *American Espionage*, 42–48; Grose, *Gentleman Spy*, 314–22.

39. Bernstein, "The Challenges and Dangers," 78–81; May, *American Cold War Strategy*.

40. Kaplan, *Wizards*, 56–58; Schwiebert, *A History of the U.S. Air Force Ballistic Missiles*, 42–44; Gorn, *Prophesy Fulfilled*.

41. "Index of ICBM Development," AWF, Admin. Series, box 17, file Guided Missiles 1958; York, *Arms*, 8–10; Kaplan, *Wizards*, 74–80, 85–110. See also Schilling, "The H-Bomb Decision"; Rhodes, *Dark Sun*, 250–53, 298, 394, 397–98, 400, 418–19.

42. Ambrose, *Eisenhower: The President*, 257.

43. "The President's Appointments, 27 March 1954," AWF, A. W. Diary Series, box 1, file ACW Diary, March 1954 (1). See also Peebles, *Corona Project*, 18–19; Hall, "The Eisenhower Administration," 62–63.

44. Immerman, "Confessions," 331; Day et al., *Eye in the Sky*, 29; Cook, *Declassified Eisenhower*, 164.

45. "Report to the National Security Council by the Special Evaluation Subcommittee of the National Security Council," NSC 140/1, May 18, 1953, in *FRUS 1952–1954: National Security Affairs*, vol. 2, pt. 1, 332–34; A NSC 140/1, Summary Evaluation of the Net Capabilities of the USSR to Inflict Direct Injury on the United States up to July 1, 1955, May 18, 1953, WHO, OSANSA, NSC Series, Policy Papers Subseries, box 3, file NSC 140/1—Special Evaluation Subcommittee of the NSC.

46. "U.S. Policy on Control of Armaments (NSC 112)," AWF, NSC Series, box 6, file 236th Meeting of NSC, February 10, 1955.

47. U.S. "Atoms for Peace" proposal, address by President Eisenhower to the General Assembly, December 8, 1953, in U.S. Department of State, *Documents on Disarmament*, document 92: 396.

48. Ambrose, *Ike's Spies*, xi; Peebles, *Shadow Flights*, 3–4; Hall, "The Eisenhower Administration," 61.

49. For a discussion of the origins of the NIE process, see Freedman, *U.S. Intelligence*, 30–31; Steury, *Sherman Kent and the Board of National Estimates*, vii, x–xvi.

50. Steury, *Intentions and Capabilities*, viii; Freedman, *U.S. Intelligence*, 31–32.

51. Letter, Brig. Gen. W. M. Burgess, deputy chief of staff/intelligence, to Brig. Gen. W. M. Garland, Air Technical Intelligence Center, May 26, 1953, SSB/GCSWS; "Briefing on Significant World Developments Affecting U.S. Security," February 17, 1954, AWF, NSC Series, box 5, file 185th Meeting of NSC, February 17, 1954.

52. Memorandum from HQ Air Defense Command to director of intelligence, Headquarters USAF, July 17, 1953, SSB/GCSWS.

53. By the time of its release the SNIE was already dated. In June 1953 the CIA increased the number of TU-4 bombers on hand to 1,600 based on a projected higher production rate of thirty-five planes per month. See "Memorandum for: Executive Secretary, NSC, Subject: CIA Comments on NSC 140/1," June 1, 1953, WHO, OSANSA, NSC Series, Policy Papers Subseries, box 3, file NSC 140/1—Special Evaluation Subcommittee of the NSC.

54. "Soviet Capabilities for Attack on the United States through Mid 1955," July 31, 1953, CIA Research Reports: The Soviet Union 1946–1976, HIL, microfilm, reel 2, frame 0649, pp. 1–3; "Appendix A: Elements of the World Situation and Outlook, the Soviet Threat through Mid-1959, NSC 5422/1," July 26, 1954, WHO, OSANSA, NSC Series, Policy Papers Subseries, box 11, file NSC 5422/2—Guidelines under NSC 162/2 for FY 1956 (2); "Memorandum of Discussion at the 157th Meeting of the National Security Council," July 30, 1953, AWF, NSC Series, box 4, file 157th Meeting of NSC, July 30, 1953.

55. "Soviet Capabilities for Attack on the United States through Mid 1955," July 31, 1953, microfilm, reel 2, frame 0649, pp. 4–5.

56. Ambrose, *Ike's Spies*, 253; Herken, *Cardinal Choices*, 87.

57. "Report to the President of the United States by the Chief of Staff, USAF, Subject: Visit of the U.S. Air Delegation to the USSR, 23 June–1 July 1956," AFW, Admin. Series, box 1, file Air Force, Department of (1).

58. Estimates of Soviet aircraft production rates could not be verified by any means available. These numbers were highly subjective and easily manipulated. A May 1953 comparison of U.S. and Soviet production of aircraft shows marginal differences in overall production whether examining the current predicted rate or an all-out production effort. See "Memorandum for Secretary Wilson," May 26, 1953, AWF, Admin. Series, box 1, file Aircraft Power.

59. "Discussion at the 172nd Meeting of the National Security Council," November 23, 1953, AWF, NSC Series, box 5, file 172nd Meeting of NSC, November 23, 1953.

60. Aronsen, "Seeing Red," 113–14, 118–19, 121–23.

61. "Memorandum for the Record," October 3, 1956, WHO, OSS, SUB, ALPHA, box 14, file Intelligence Matters (2); "The Reminiscences of Andrew J. Goodpaster," April 10, 1982, 18–20, Dwight D. Eisenhower Presidential Library.

62. Kaplan, *Wizards*, 85–110; Richelson, *America's Secret Eyes in Space*, 9–10.

63. "Significant World Developments Affecting U.S. Security," April 29, 1954, AWF, NSC Series, box 5, file 194th Meeting of NSC, April 29, 1954; "Briefing on Significant World Developments Affecting U.S. Security," May 13, 1954, AWF, NSC Series, box 5, file 197th Meeting of NSC, May 13, 1954; "Significant World Developments

Affecting U.S. Security," April 28, 1955, AWF, NSC Series, box 6, file 246th Meeting of NSC, April 28, 1955; NIE 11-4-54, "Soviet Capabilities & Probable Courses of Action through Mid 1959," September 14, 1954, in Steury, *Estimates*, 1–37.

64. NIE 11-5-54, "Soviet Capabilities and Main Lines of Policy through Mid-1959," in Koch, *Selected Estimates*, 201–11. See also memorandum by the acting special assistant to the secretary of state for intelligence (Howe) to the acting secretary of state, subject: SNIE 11-2-54: Soviet Capabilities for Attack on the U.S. through 1957, March 1, 1954, in FRUS: *National Security Affairs*, vol. 2, pt. 1, 634–737.

65. For example, the White House received telegrams and phone calls pushing for rapid increases in spending on air power. See "Telegram from Congressman Samuel W. Yorty (Republican, California) to DDE," October 4, 1953, AWF, A. W. Diary Series, box 3, file ACW Diary, October 1954 (5).

66. The figures for these bomber numbers come from the Natural Resources Defense Council, which has published a great deal of information regarding nuclear force size since 1984. Two of the most important working papers dealing with nuclear data are "U.S.-USSR/Russian Strategic Offensive Nuclear Forces 1945–1996" (January 1997) and "U.S. Inventories of Nuclear Weapons and Weapon-Usable Fissile Material" (revised September 26, 1995), Natural Resources Defense Council, "Archive of Nuclear Data," http://www.nrdc.org/nuclear/nudb/datab7.asp#foot1.

67. NIE 11-3-55 (May 1955) and NIE 11-56 (March 1956) predicted forty Bison bombers by January 1, 1956, and eighty by July 1, 1956. In NIE 11-4-56 (August 1956) this number was dropped to thirty-five. The decrease was due to slower Bison production. Observations of an aircraft plant in Moscow and new data on Soviet long range aviation bases supported this. See "Memorandum of Discussion at the 280th Meeting of the National Security Council," March 22, 1956, AWF, NSC Series, box 7, file 280th Meeting of NSC, March 22, 1956; "Memorandum for Brig. General Goodpaster from DCI Dulles," March 1, 1957, WHO, OSS, SUB, ALPHA, box 14, file Intelligence Matters (3); "Memorandum for the Record," February 12, 1959, WHO, OSS, SUB, ALPHA, box 15, file Intelligence Matters (8); Freedman, *U.S. Intelligence*, 66–67; Peebles, *Shadow Flights*, 136–37, 149.

68. "Soviet Gross Capabilities for Attack on the U.S. and Key Overseas Installations and Forces through Mid-1959 (NIE 11-56)," in Steury, *Intentions and Capabilities*, 5–6, 9–19; "Memorandum of Discussion at the 409th Meeting of the National Security Council," June 4, 1959, AWF, NSC Series, box 11, file 409th Meeting of NSC, June 4, 1959; "Memorandum for the President, Subject: Supplemental Appropriations for FY '57," March 29, 1956, AWF, Admin. Series, box 9, file Budget 1957 (2); diary entry for April 4, 1956, and "Memorandum for the Record," April 4, 1956, AWF, A. W. Diary Series, box 8, file Apr. '56 Diary-ACW (2); diary entry for April 26 and 27, 1956, AWF, A. W. Diary Series, box 8, file Apr. '56 Diary-ACW (1); "Address by General Thomas D. White, Vice Chief of Staff, USAF to the General Electric Dinner," February 9, 1956, WHO, OSS, SUB, ALPHA, box 6, file Military Program (Missiles) [January 1956–September 1957] (2).

69. SNIE 11-6-57, "Soviet Gross Capabilities for Attacks on the Continental U.S. in Mid-1960," was produced in October 1956 (in Steury, *Intentions and Capabilities*, 39–46, 47–56). See also Koch, *Selected Estimates*, 215–21; Freedman, *U.S. Intelligence*, 21–22; Herken, *Cardinal Choices*, 87.

70. "Index of ICBM Development," AWF, Admin. Series, box 17, file Guided Missiles 1958; Schwiebert, *History*, 42–46; Neufeld, *Development*, 44–48.

71. "Summary Presentation on History of Development of U.S. Long-Range Guided Missiles by Herbert F. York, Director of Defense Research and Engineering," May 5, 1960, WHO, NSCSP, Exec. Sect. Subject File Series, box 1, file Miscellaneous (File #2) (7); Kistiakowsky, *A Scientist*, 95–96; Levine, *Missile*, 29–34; Killian, *Sputniks*, 11–13; Schwiebert, *History*, 70–78.

72. Diary entry, April 4, 1956, AWF, A. W. Diary Series, box 8, file Apr. '56 Diary-ACW (2); diary entry, April 27, 1956, AWF, A. W. Diary Series, box 8, file Apr. '56 Diary-ACW (1); "Meet the Press, Transcript for Sunday 5 February 1956," WHO, OSS, SUB, DoD Subseries, box 6, file Military program (Missiles) [January 1956–September 1957] (2); "Address by General Thomas D. White, Vice Chief of Staff, USAF at the General Electric Dinner," February 9, 1956, WHO, OSS, SUB, DoD Subseries, box 6, file Military program (Missiles) [January 1956–September 1957] (2).

73. "Memorandum of Discussion at the Special Meeting of the National Security Council," March 31, 1953, AWF, NSC Series, box 4, file Special Meeting of NSC, March 31, 1953; "Memorandum of Discussion at the Special Meeting of the National Security Council," March 31, 1953, in *FRUS: National Security Affairs*, vol. 2, pt. 1, 268.

74. "Continental Defense, NSC 5408," February 11, 1954, WHO, OSANSA, NSC Series, Policy Papers Subseries, box 9, file NSC 5408 Continental Defense (1).

75. Gaddis, "Transparency," 197; Hall, "Origins of U.S. Space Policy," 218–19; Burrows, *This New Ocean*, 155; Taubman, *Secret Empire*, 24–25.

76. These concerns were not new. On June 4, 1953, President Eisenhower raised his concerns over not being able to penetrate the Soviet Union to gain any useful intelligence. See "Discussion at the 148th Meeting of the National Security Council," June 4, 1953, AWF, NSC Series, box 4, file 148th Meeting of NSC, June 4, 1953.

77. McElheny, *Insisting*, 278–79; Ambrose, *Ike's Spies*, 267–68; Burrows, *This New Ocean*, 152–59. Peebles, *Corona Project*, 18–20; Killian, *Sputniks*, 11–13.

78. Ferris, "Coming in from the Cold War," 90.

2. Eisenhower and Defense

"Farewell Address by President Eisenhower on the 'Military Industrial Complex,'" in LaFeber, *Dynamics*, 645–46.

1. Metz, "Eisenhower," 51.

2. Metz, "Eisenhower," 51.

3. Presidential News Conference, May 14, 1953, in Branyan and Larsen, *Eisenhower Administration*, 1:39; Taubman, *Secret Empire*, xii.

4. "Informal Condensation of NSC 20/4, 68/2, 135/3, and 141 (for discussion purposes at NSC meetings)," February 6, 1953, WHO, OSANSA, NSC Series, Subject Subseries, box 8, file President's Meeting with Civilian Consultants, March 31, 1953 [re: Review of Basic National Security Policy] (1).

5. May, *American Cold War Strategy*, 25.

6. May, *American Cold War Strategy*, 23–80. See also Leffler, *Preponderance*, 355–60; Bernstein, "The Challenges and Dangers," 79–80.

7. These views continued in NSC 135/3 (September 25, 1952) with no major changes. "Informal Condensation of NSC 20/4, 68/2, 135/3, and 141 (for discussion purposes at NSC meetings)"; May, *American Cold War Strategy*, 23–80; Leffler, *Preponderance*, 355–60; "The Reminiscences of Andrew J. Goodpaster," April 10, 1982, 1–3.

8. May, *American Cold War Strategy*, vii; Robert J. Donovan, "Truman's Perspective," in Heller, *Economics*, 18–19; John F. Snyder, "The Treasury and Economic Policy," in Heller, *Economics*, 29–30.

9. Ambrose, *Eisenhower: The President*, 88; Donovan, *Eisenhower*, 51; Bernstein, "The Challenges and Dangers," 80–81.

10. DDE Personal Diary, January 22, 1952, box 1, file [January 1–February 28, 1952], 3. See also "Letter to Charles E. Wilson from President Eisenhower, Jan. 5, 1955," AWF, DDE Diary Series, box 9, file DDE Diary, January 1955 (2).

11. "Statement of Policy, General," April 29, 1953, WHO, OSANSA, NSC Series, Policy Papers, box 4, file NSC 149-2—Basic National Security Policies & Programs in Relation to Their Costs (1); "The Reminiscences of Andrew J. Goodpaster," April 10, 1982, 1–3; "Press Release of an Exchange of Correspondence between the President and Secretary of Defense, 5 January 1955," WHO, OSS, SUB, DoD Subseries, box 6, file Military Planning, 1954–1955 (1).

12. DDE Personal Diary, January 22, 1952, box 1, file [January 1, 1950–February 28, 1952], 2; DDE Personal Diary, June 4, 1949, box 1, file [December 13, 1948–March 5, 1951], 1; Darrell B. Montgomery, "New Evidence of the Evolution of a Postwar Air-Atomic Strategy," in Levantrosser, *Harry S. Truman*, 158–59.

13. DDE Personal Diary, December 17, 1948, January 8, 1949, June 4, 1949, box 1, file [December 13, 1948–March 5, 1951], 1; January 22 1952, box 1, file [January 1, 1950–February 28, 1952], 2; "Notes on Legislative Leadership Meeting, 30 April 1953," AWF, DDE Diary Series, box 4, file Staff Notes January–December 1953.

14. "Notes on Legislative Leadership Meeting, 30 April 1953"; Ambrose, *Eisenhower: The President*, 88–89.

15. Diary entry, December 13, 1954, Hagerty Diary, box 1A, file December 1954; letter to Charles E. Wilson from President Eisenhower, January 5, 1955, AWF, DDE Diary Series, box 9, file DDE Diary, January 1955 (2).

16. Letter to Edward Hazlett from General Eisenhower, Hazlett Papers, box 1, file 1946 July 1.

17. Letter to John Foster Dulles and Neil McElroy from Robert Cutler, April 7, 1958, WHO, OSANSA, NSC Series, Briefing Notes Subseries, box 14, file Nuclear Policy (1958).

18. "Memorandum for Admiral Arthur W. Radford, USN, Chairman, Joint Chiefs of Staff, 17 February 1956," WHO, OSS, SUB, DoD, box 4, file Joint Chiefs of Staff (2) [January–April 1956]; "The Reminiscences of Andrew J. Goodpaster," OH-378, 104–5.

19. Diary entry, June 19, 1954, Hagerty Diary, box 1, file June 1954. See also "The Next Ten Years," WHO, OSANSA, NSC Series, Policy Papers Subseries, box 2, file NSC 112/1—Disarmament (3).

20. Diary entry, February 1, 1955, Hagerty Diary, box 1A, file February 1955.

21. "Memorandum: Discussion at the 272nd Meeting of the National Security Council, January 12, 1956," AWF, NSC Series, box 7, file 272nd Meeting of NSC, January 12, 1956.

22. "Memorandum: Discussion at the 272nd Meeting of the National Security Council, January 12, 1956." See also memorandum of conference with the president, May 24, 1956, AWF, DDE Diary Series, box 15, file May '56 Goodpaster.

23. The worst-case scenario called for an attack without any warning until U.S. radar detected the bombers. The "better"-case scenario allowed the United States to have approximately one month's warning, though no information on the date of such an attack. See "Memorandum: Discussion at the 201st Meeting of the National Security Council, June 10, 1954," AWF, NSC Series, box 5, file 201st Meeting of NSC, June 10, 1954; diary entry, January 23, 1956, AWF, DDE Diary Series, box 9, file Diary—Copies of DDE Personal [1955–56] (2); "Report to the NSC by the Special Evaluation Subcommittee of the NSC," May 18, 1953, in *FRUS 1952–1954: National Security Affairs*, vol. 2, pt. 1, 329–35.

24. "Report by the Net Evaluation Subcommittee," December 20, 1956, AWF, NSC Series, box 8, file 306th Meeting of NSC, December 20, 1956; "Report by the Net Evaluation Subcommittee," November 12, 1957, AWF, NSC Series, box 9, file 344th Meeting of NSC, November 12, 1957; "Memorandum of Conference with the President, 4 November 1957," WHO, OSS, SUB, ALPHA, box 23, file Science Advisory Committee (3).

25. Letter from Eisenhower to Richard L. Simon, April 4, 1956, AWF, DDE Diary Series, box 14, file April 1956 Miscellaneous (5). See also "Legislative Leadership Meeting, Supplementary Notes, 24 June 1958," AWF, DDE Diary Series, box 33, file June 1958—Staff Notes (2); "Press Release of an Exchange of Correspondence between the President and Secretary of Defense, 5 January 1955," WHO, OSS, SUB, DoD, box 6, file Military Planning, 1954–1955 (1).

26. "Guidelines under NSC 162/2 for FY 1956," July 26, 1954, WHO, OSANSA, NSC Series, Policy Papers Subseries, box 11, file NSC 5422/2—Guidelines under 162/2 for FY 1956 (2); memorandum for the secretary of state from President Eisenhower, September 8, 1953, AWF, DDE Diary Series, box 3, file DDE Diary—August–September 53 (2); memorandum: "Subject: The Meaning of Paragraph 39b, NSC 162/2, as Understood by the Department of Defense," December 1, 1953, WHO, NSCSP, Exec. Sect. Subject File Series, box 5, file #19 Policy re Use (of nuclear weapons) (file #1) (1).

27. "Letter to General Alfred M. Gruenther, Chief of Staff, SHAPE, from President Eisenhower, 4 May 1953," AWF, DDE Diary Series, box 3, file DDE Diary, Decem-

ber 1952–July 1953 (3); "Bipartisan Legislative Meeting, 5 January 1954," AWF, DDE Diary Series, box 4, file Staff Notes, January–December 1954.

28. Memorandum to the NSC by Executive Secretary Lay, "Review of Basic NSC Policies," February 6, 1953, in *FRUS: National Security Affairs*, vol. 2, pt. 1, 223–24; Freedman, *Evolution*, 81–82; "Notes by the Assistant Staff Secretary to the President (Minnich) on the Legislative Leadership Meeting," December 14, 1954, in *FRUS: National Security Affairs*, vol. 2, pt. 1, 826; *Congress and the Nation*, 274–75; "Basic National Security Policy," January 7, 1955, WHO, OSANSA, NSC Series, Policy Papers Subseries, box 14, file NSC 5501—Basic National Security Policy.

29. "Annex A," WHO, NSCSP, Exec. Sect. Subject File Series, box 11, file General Papers (Colonel Bonesteel); "The Reminiscences of Andrew J. Goodpaster," April 10, 1982, 13; "The Reminiscences of Andrew J. Goodpaster," April 25, 1967, 12–14, Columbia Center for Oral History.

30. Memorandum for the record by the special assistant to the president for national security affairs (Cutler), "SOLARIUM Project," May 9, 1953, in *FRUS: National Security Affairs*, vol. 2, pt. 1, 323–26; memorandum for the National Security Council, "Project Solarium," July 22, 1953, WHO, OSANSA, NSC Series, Subject Subseries, box 9, file Project Solarium, Report to the NSC by Task Force "A"; "Project Solarium: Summary of Basic Concepts of Task Forces," July 30, 1953, WHO, OSANSA, NSC Series, Subject Subseries, box 10, file Project Solarium [1953] (1).

31. Memorandum for the National Security Council, "Project Solarium," July 22, 1953.

32. "Annex A"; memorandum for the National Security Council, "Project Solarium," July 22, 1953; "Task Force B," July 16, 1953, WHO, OSANSA, NSC Series, Subject Subseries, box 9, file Project Solarium, Report to the NSC by Task Force "A" [1953] (2); "The Reminiscences of Andrew J. Goodpaster," April 10, 1982, 13.

33. "Annex A."

34. "Task Force C," July 16, 1953, WHO, OSANSA, NSC Series, Subject Subseries, box 9, file Project Solarium, Report to the NSC by Task Force "A" [1953] (2); memorandum for the National Security Council, "Project Solarium," July 22, 1953; "The Reminiscences of Andrew J. Goodpaster," April 10, 1982, 13.

35. "Project Solarium," July 16, 1953, AWF, NSC Series, box 4, file Minutes of 155th Meeting of NSC, July 16, 1953; "The Reminiscences of Andrew J. Goodpaster," April 10, 1982, 14–15.

36. "The Reminiscences of Andrew J. Goodpaster," April 10, 1982, 14–15.

37. Gaddis, *Strategies*, 133.

38. "Meeting with Secretary Dulles, 7/20/54," AWF, A. W. Diary Series, box 2, file ACW Diary, July 1954 (3); Freedman, *Evolution*, 76–78; Kinnard, *Secretary of Defense*, 51–52; Bernstein, "The Challenges and Dangers," 82–83.

39. Diary entry, December 9 and December 13, 1954, Hagerty Diary, box 1A, file December 1954; "Memorandum of Conference with the President, 6 November 1957," WHO, OSS, SUB, ALPHA, box 23, file Science Advisory Committee (3); "Memorandum for the National Security Council, Subject: Basic National Secu-

rity Policy, 31 March 1960," WHO, NSCSP, Exec. Sect. Subject File Series, box 13, file NSC 5906/1—Basic National Security Policy [1959]; memorandum of conference with the president, May 24, 1956, AWF, DDE Diary Series, box 15, file May 1956; "The Reminiscences of Andrew J. Goodpaster," April 10, 1982, 14–15; Freedman, *Evolution*, 76–78; Kinnard, *Secretary of Defense*, 51–52; Bernstein, "The Challenges and Dangers," 82–83.

40. All amounts, except percentages, are in U.S. dollars; figures from U.S. Bureau of the Census, *Historical Statistics of the United States: Colonial Times to 1970*, 1975, pt. 1, 224, 230, 233; pt. 2, 1116, https://fraser.stlouisfed.org/title/?id=237, accessed July 15, 2015; Divine, *Since 1945*, 7–30; Craufurd A. Goodwin, "The Economic Problems Facing Truman," in Heller, *Economics*, xv–xvii, 1–2, 13–20; Gosnell, *Truman's Crises*, 257–390.

41. All amounts, except percentages, are in U.S. dollars; figures from *Historical Statistics of the U.S., Colonial Times to 1970*, pt. 1, 224, 230, 233; pt. 2, 1116; letter to Eisenhower from Secretary Humphrey, July 29, 1953, AWF, DDE Diary Series, box 3, file DDE Diary December 1952–July 1953 (1); notes on legislative leadership meeting, April 30, 1953, AWF, DDE Diary Series, box 4, file Stuff Notes January–December 1953; Divine, *Since 1945*, 7–30; Goodwin, "The Economic Problems Facing Truman," xv–xvii, 1–2, 13–20; Gosnell, *Truman's Crises*, 257–390.

42. Harris, *Economics*, xv–xxiii, 3–8; Heller, *Economics*, 6–7; letter to Clifford Roberts from Eisenhower, March 27, 1958, AWF, DDE Diary Series, box 31, file DDE Dictation, March 1958.

43. Speech before the American Legion Convention, New York City, August 25, 1952, in Eisenhower Library, "Campaign Statements of Dwight D. Eisenhower: A Reference Guide," 111.

44. Campaign speeches in Philadelphia, September 4, 1952, Jackson, Michigan, October 1, 1952, New Brunswick, New Jersey, October 17, 1952, in Eisenhower Library, "Campaign Statements of Dwight D. Eisenhower: A Reference Guide," 111.

45. Letter to Eisenhower from G. M. Humphrey, July 29, 1953, AWF, DDE Diary Series, box 3, file DDE Diary December 1952–July 1953 (1); letter to Brig. Gen. Benjamin F. Caffey (Ret.) from Eisenhower, July 27, 1953, AWF, DDE Diary Series, box 3, file DDE Diary December 1952–July 1953 (1).

46. "Memorandum of Discussion at the 138th Meeting of the National Security Council," March 25, 1953, in *FRUS: National Security Affairs*, vol. 2, pt. 1, 262–63; letter to Brig. Gen. Benjamin F. Caffey (Ret.) from DDE, July 27, 1953; letter to Swede Hazlett, July 22, 1957, AWF, DDE Diary Series, box 25, file July 1957—DDE Dictation; Ambrose, *Ike's Spies*, 275–76.

47. "Review of the 1954 Budget," August 27, 1953, AWF, Admin. Series, box 12, file Dodge, Joseph M. 1952–1953 (1).

48. Discussion at the 165th meeting of the National Security Council, October 7, 1953, AWF, NSC Series, box 4, file 165th Meeting of NSC, October 7, 1953; "Condensed Statement of Proposed Policies and Programs (Draft), 31 March 1953," AWF, NSC Series, box 4, file Documents Pertaining to Special NSC Meeting, March 31, 1953.

49. Memorandum of conference with the president, April 2, 1956, WHO, OSS, SUB, DoD Subseries, box 4, file Joint Chiefs of Staff (2) [January–April 1956]; memorandum of conference with the president, November 6, 1957, WHO, OSS, SUB, ALPHA, box 23, file Science Advisory Committee (3); undated statement by Eisenhower, AWF, A. W. Diary Series, box 8, file January '57 Diary-ACW.

50. "Review of Basic National Security Policy, NSC 162," October 7, 1953, AWF, NSC Series, box 4, file 165th Meeting of NSC, October 7, 1953; diary entry December 14, 1954, Hagerty Diary, box 1A, file December 1954; bipartisan legislative meeting, December 14, 1954, AWF, DDE Diary Series, box 4, file Staff Notes, January–December 1954.

51. DDE Personal Diary, January 8, 1949, box 1, file [DDE Diary, December 13, 1948–March 5, 1951], 1; Divine, *Since 1945*, 7–8, 28–30; Montgomery, "New Evidence," 158–60.

52. Eisenhower left Korea on December 5, 1952, flying to Guam, where he boarded the cruiser USS *Helena*. From there he sailed to Wake, where members of his future cabinet met the ship. Dulles, Humphrey, McKay, General Clay, and Dodge joined up with the president-elect to discuss a variety of matters. See Eisenhower, *White House Years*, 1:96; Ambrose, *Eisenhower: The President*, 30–31.

53. "Capsule Statement of National Security Policy," AWF, NSC Series, box 4, file Documents Pertaining to Special NSC Meeting, March 31, 1953; letter to Clifford Roberts from Eisenhower, March 27, 1958, AWF, DDE Diary Series, box 31, file DDE Dictation, March 1958; Ambrose, *Eisenhower: The President*, 32–34; Cook, *Declassified Eisenhower*, 300; Freedman, *Evolution*, 156.

54. "State of the Union Address," February 2, 1953, in *Public Papers*, 1:17.

55. "Draft Statement of Policy Proposed by the National Security Council," February 11, 1954, in *FRUS: National Security Affairs*, vol. 2, pt. 1, 611. For similar statements, see "Restatement of Basic National Security Policy, General Considerations [DRAFT STATEMENT]," June 1, 1953, WHO, OSANSA, NSC Series, Policy Papers Subseries, box 5, file NSC 153/1—Basic National Security Policy.

56. Notes on meeting of April 4, 1956, AWF, A. W. Diary Series, box 8, file April '56 Diary-ACW (2).

57. *Congress and the Nation*, 274. See also memorandum of conference with the president, September 30, 1958, AWF, DDE Diary Series, box 36, file Staff Notes—September 1958.

58. Kistiakowsky, *A Scientist*, xviii–xix.

59. Radio address to the American people on National Security and its costs, May 19, 1953, in *Public Papers*, 1:306–16; "Condensed Statement of Proposed Policies and Programs (Draft Statement)," April 2, 1953, AWF, NSC Series, box 4, file Documents Pertaining to Special NSC Meeting, March 31, 1953; DDE Personal Diary, January 22, 1955, box 1, file [January 1, 1950–February 28, 1952], 3.

60. *Congress and the Nation*, 275–76.

61. "Memorandum of Conference with the President," December 22, 1954, AWF, A. W. Diary Series, box 3, file ACW Diary, December 1954 (2); DDE Personal Diary,

January 22, 1952, box 1, file [January 1, 1950–February 28, 1952], 2; diary entry, December 13–14, 1954, Hagerty Diary, box 1A, file December 1954; president's news conference, April 30, 1953, in *Public Papers*, 1:245–46.

62. "Notes on Legislative Leadership Meeting, 30 April 1953," AWF, DDE Diary Series, box 4, file Staff Notes January–December 1953.

63. Ambrose, *Eisenhower: The President*, 88–89; Gaddis, *Strategies*, 139–40. See also diary entry, December 13, 1954, Hagerty Diary, box 1A, file December 1954.

64. Memorandum of conference with the president, December 20, 1956, WHO, OSS, SUB, DoD Subseries, box 2, file Budget, Military (4) [May–December 1956]; discussion at the 293rd and 294th meetings of the National Security Council, August 16 and 17, 1956, AWF, NSC Series, box 8, file 293rd and 294th Meetings of NSC, August 16 and 17, 1956; Gaddis, *Strategies*, 133.

65. Eisenhower, as quoted in Gaddis, *Strategies*, 133–34. Andrew Goodpaster maintains that this speech clearly showed the president's commitment to both disarmament and the control of defense spending. Author's phone interview with Goodpaster, March 7, 2003.

66. Gaddis, *Strategies*, 133–34; Ambrose, *Eisenhower: The President*, 224; Eisenhower, *White House Years*, 1:144–45; Divine, *Eisenhower and the Cold War*, 108.

67. "Notes on Legislative Leadership Meeting," April 30, 1953; diary entry, June 1, 1953, DDE Personal Diary, box 1, file December 1952–8/19/53 (2); discussion of the 138th meeting of the National Security Council, March 25, 1953, AWF, NSC Series, box 4, file 138th Meeting of NSC, March 25, 1953.

68. "Memorandum for the Director of the Bureau of the Budget from DDE," December 1, 1953, AWF, Admin. Series, box 12, file Dodge, Joseph M. 1955 Budget (2); "Review of National Security Programs, Memorandum for the Executive Secretary, National Security Council," March 24, 1953, WHO, OSANSA, NSC Series, Subject Subseries, box 8, file President's Meeting with Civilian Consultants, March 31, 1953 [re: Review of Basic National Security Policy] (8).

69. Joseph Alsop and Stewart Alsop, "Defense 'New Look' and New Weapons," *Washington Post*, February 22, 1954, AWF, NSC Series, box 5, file 187th Meeting of NSC, March 4, 1954; "The Reminiscences of General Nathan F. Twining," 142–44; Presidential News Conference, March 17, 1954, in Branyan and Larsen, *Eisenhower Administration*, 1:40–41.

70. Discussion at the 160th meeting of the National Security Council, August 27, 1953, AWF, NSC Series, box 4, file 160th Meeting of NSC, August 27, 1953.

71. Discussion at the 166th meeting of the National Security Council, October 13, 1953, AWF, NSC Series, box 4, file 166th Meeting of NSC, October 13, 1953; Alsop and Alsop, "Defense 'New Look'"; draft memorandum, "Subject: Policy Regarding Use of Nuclear Weapons," December 31, 1953, signed by Eisenhower January 2, 1954; memorandum "Subject: The Meaning of Paragraph 39b, NSC 162/2, as Understood by the Department of Defense," December 1, 1953; memorandum for the record from Robert Cutler, December 2, 1953, WHO, NSCSP, Exec. Sect. Subject File Series, box 5, file #19 Policy re Use (of Nuclear Weapon) (File # 1) (1).

72. *Congress and the Nation*, 275; diary entry, December 13, 1954, Hagerty Diary, box 1A, file December 1954; "Principal Budgetary Elements of New Obligational Authority and Expenditures," May 27, 1953, AWF, Admin. Series, box 12, file Dodge, Joseph M. 1952–1953 (1); Eisenhower, *White House Years*, 1:452.

73. "Budget Message of the President," January 12, 1954, AWF, Admin. Series, box 12, file Dodge, Joseph M. 1954–56 (5); *Congress and the Nation*, 280; Ambrose, *Eisenhower: The President*, 223; Eisenhower, *White House Years*, 1:452.

74. *Congress and the Nation*, 275, 279–80, 285–92, 294–307; Department of Defense Military New Obligational Authority, undated, AWF, Admin. Series, box 8, file Brundage, Percival 1955–57 (2).

75. Letter to Secretary of Defense Charles E. Wilson from Eisenhower, January 5, 1955, WHO, OSS, SUB, DoD Subseries, box 5, file Man Power and Personnel (2) [January 1955–August 1957]; Ambrose, *Eisenhower: The President*, 144, 171–72; Ambrose, *Rise to Globalism*, 135–36.

76. "President's Interview with John Taber, Congressman from New York," October 21, 1953, AWF, A. W. Diary Series, box 1, file ACW Diary, August–September–October 1953 (1).

77. Exchange of correspondence between Secretary Wilson and Eisenhower, January 5, 1955, WHO, OSS, SUB, DoD Subseries, box 6, file Military Planning, 1954–1955 (1); *Congress and the Nation*, 274–75, 279–80.

78. Letter to George Humphrey from Eisenhower, November 22, 1957, AWF, DDE Diary Series, box 28, file November '57 DDE Diary.

79. Memorandum of conference with the president, December 20, 1956, WHO, OSS, SUB, DoD Subseries, box 2, file Budget, Military (4) [May–December 1952]; memorandum of conference with the president, April 2, 1956, WHO, OSS, SUB, DoD Subseries, box 4, file Joint Chiefs of Staff (2) [January–April 1956]; diary entry, February 1, 1955, Hagerty Diary, box 1A, file February 1955.

80. DDE Personal Diary, December 17, 1948, January 27, 1949, box 1, file 1948, 1; letter to Swede Hazlett from Eisenhower, April 27, 1949, Hazlett Papers, box 1, file 1949 April 27.

81. DDE Personal Diary, January 7, 1949, box 13, file 1948 (Washington).

82. DDE Personal Diary, January 7–8, 1949, February 2, 1949, box 13, file 1948 (Washington).

83. DDE Personal Diary, February 2 and 4, 1949, box 13, file 1948 (Washington); letter to Swede Hazlett, February 24, 1950, box 1, file 1950 February 24; letter to Swede Hazlett, November 14, 1951, box 1, file 1951 November 14.

84. DDE Personal Diary, January 22, 1952, box 1, file [January 1, 1950–February 28, 1952], 3; "Discussion at the 163rd Meeting of the National Security Council," September 24, 1953, AWF, NSC Series, box 4, file 163rd Meeting of NSC, September 24, 1953.

85. An excellent example was the open journal discussions of new aircraft capabilities, which only aggravated competition between the services. See "Memorandum of Conference with the President," April 18, 1956, AWF, DDE Diary Series,

box 15, file Apr. '56 Goodpaster; "Memorandum of Conference with the President," October 7, 1958, AWF, DDE Diary Series, box 36, file Staff Notes—October 1958; discussion at the 141st meeting of the National Security Council, May 6, 1953, AWF, NSC Series, box 4, file 143rd Meeting of NSC, May 6, 1953.

86. "Memorandum for the Record," November 6, 1957, AWF, DDE Diary Series, box 28, file November '57, Staff Notes.

87. "Memorandum for the Record," November 6, 1957, WHO, OSS, SUB, DoD Subseries, box 1, file Department of Defense, Vol. 2 (3) [November–December 1957]; "Memorandum for the Record," November 6, 1957, AWF; phone conversation with Gen. Andrew Goodpaster, March 7, 2003.

88. "Discussion on Continental Defense," November 23, 1953, AWF, NSC Series, box 5, file 172nd Meeting of NSC, November 23, 1953; "Memorandum of Conference with the President," December 6, 1954, AWF, A. W. Diary Series, box 3, file ACW Diary, 1954 (5).

89. Diary entry, October 26, 1953, AWF, A. W. Diary Series, box 1, file ACW Diary, August–September–October 1953 (1); "Discussion at the 293rd and 294th Meetings of the National Security Council," August 16 and 17, 1956, AWF, NSC Series, box 8, file 293rd and 294th Meeting of NSC, August 16 and 17, 1956; "The Reminiscences of General Nathan F. Twining," 113–17; "Meeting at the Pentagon," January 25, 1958, WHO, OSS, SUB, DoD Subseries, box 1, file Department of Defense, Vol. 2 (4) [January 1958].

90. Diary entry, December 9, 1954, Hagerty Diary, box 1A, file December 1954; "Discussion at the 138th Meeting of the National Security Council," March 25, 1953, AWF, NSC Series, box 4, file 138th Meeting of NSC, March 25, 1953.

91. Diary entry, February 1, 1955, "Legislative Leaders Meeting," Hagerty Diary, box 1A, file February 1955; memorandum of conference with the president, April 5, 1956, AWF, DDE Diary Series, box 15, file Apr. '56 Goodpaster; diary entry, December 9, 1954, Hagerty Diary, box 1A, file December 1954; discussion at the 176th meeting of the National Security Council, December 16, 1953, AWF, NSC Series, box 5, file 176th Meeting of NSC, December 16, 1953; discussion at the 227th meeting of the National Security Council, December 3, 1954, AWF, NSC Series, box 6, file 227th Meeting of NSC, December 3, 1954.

92. "The Reminiscences of John S. D. Eisenhower," 14–38; "The Reminiscences of General Nathan F. Twining," 113–17, 144–45.

93. Ambrose, *Eisenhower: Soldier and President*, 321–22; *Congress and the Nation*, 279–80.

94. Notes on legislative leadership meeting, April 30, 1953, AWF, DDE Diary Series, box 4, file Staff Notes January–December 1953.

95. Ambrose, *Eisenhower: The President*, 88.

96. "Cardboard Wings vs Real Strength," June 1953, Harlow Records, box 4, file Air Force Appropriations [1953 and 1957–58] (4); "Salient Facts about Air Force Expenditures," June 5, 1953, Harlow Records, box 4, file Air Force Appropriations [1953 and 1957–58] (5); "Air Force Strength," January 5, 1954, WHO, OSS, Minnich

Series, box 1, file Miscellaneous—A (2) [July 1953–February 1954]; "Special Staff Note on Military Aircraft Production," July 27, 1957, AWF, DDE Diary Series, box 25, file July 1957 Staff Memos (1).

97. Ambrose, *Eisenhower: The President*, 88–91; Ambrose, *Eisenhower: Soldier and President*, 321–22; Hall, "The Eisenhower Administration," 60–61; Divine, *Sputnik Challenge*, 20–21; "The Reminiscences of Andrew J. Goodpaster," April 10, 1982, 18; Huntington, *Changing Patterns*, 14.

98. The budget for fiscal year 1954 was not the first of the New Look; most of the groundwork and long-term planning that went into it was a carryover from the Truman administration.

99. *Congress and the Nation*, 279–89; memorandum for executive secretary, NSC, from Allen Dulles, DCI, June 1, 1953, WHO, OSANSA, NSC Series, Policy Papers Subseries, box 3, file NSC 140/1—Special Evaluation Subcommittee of the NSC.

100. Diary entry, April 27, 1956, AWF, A. W. Diary Series, box 8, file April '56 Diary-ACW (1); *Congress and the Nation*, 290–92; Freedman, *Evolution*, 29; "Pre–Press Conference Briefing," February 8, 1956, AWF, DDE Diary Series, box 13, file February '56 Miscellaneous (5).

101. Ambrose, *Eisenhower: The President*, 223–24.

102. Eisenhower et al., *Ike's Letters*, 109; diary entry, February 1, 1955, Hagerty Diary, box 1A, file February 1955.

103. Ambrose, *Eisenhower: The President*, 225.

104. Letter to the president from Senator Symington, August 29, 1958, AWF, Admin. Series, box 36, file Symington, Senator Stuart.

105. "Memorandum for the Record, Subject: Discussion of Soviet and U.S. Long Range Ballistic Missile Programs," August 18, 1958, WHO, OSS, SUB, ALPHA, box 24, file Symington Letter.

106. Letter to the president from Senator Symington, August 29, 1958; "Memorandum of Conversation, Subject: DCI Briefing of Senator Stuart Symington on *Soviet Ballistic Missile Programs and Capabilities*," December 16, 1958, WHO, OSS, SUB, ALPHA, box 24, file Symington Letter.

107. Letter from Eisenhower to Everett Hazlett, August 20, 1956, quoted in Ambrose, *Eisenhower: The President*, 225.

108. "Memorandum for the National Security Council, Subject: Basic National Security Policy, 31 March 1960," WHO, NSCSP, Exec. Sect. Subject File Series, box 13, file NSC 5906/1—Basic National Security Policy [1959]; Freedman, *Evolution*, 76–78; Kinnard, *Secretary of Defense*, 51–52.

109. Bissell et al., *Reflections*, 124–25.

3. Satellite Reconnaissance

"Comments on the Report to the President by the Technological Capabilities Panel of the Science Advisory Committee," WHO, OSANSA, NSC Series, Policy Papers Subseries, box 16, folder NSC 5522 Technological Capabilities Panel, S23.

1. Burrows, *This New Ocean*, 157–58.

2. Bissell et al., *Reflections*, 92.

3. Davies and Harris, RAND's *Role*, 55, RAND Corporation; Beschloss, *May-Day*, 73–74. Project RAND was an army air forces–sponsored think tank that used a multidisciplinary approach to examine the long-term issues and questions that the air force itself could not. Davies and Harris, RAND's *Role*, 3–6; Kaplan, *Wizards*, 1–5, 51–63.

4. The numerous RAND studies on American vulnerability and the need to improve warning include "The Cost of Decreasing Vulnerability of Air Bases by Dispersal," R-235, June 1, 1952; "The Military Value of Advanced Warning of Hostilities and Its Implications for Intelligence Indicators," July 1953; "Vulnerability of U.S. Strategic Air Power to a Surprise Attack in 1956," SM-15, April 15, 1953. See Davies and Harris, RAND's *Role*, 48–51; Kaplan, *Wizards*, 85–110, 116–27; Taubman, *Secret Empire*, 13–14.

5. Killian, *Sputniks*, 68–69; Welzenbach, "Din Land," 22; "The Reminiscences of James R. Killian," pt. 1, p. 13, pt. 8, pp. 224–25, 228–34.

6. Killian, *Sputniks*, 69–70; Welzenbach, "Din Land," 22–23; Peebles, *Corona Project*, 18; letter to Gen. Curtis E. LeMay from James R. Killian Jr., September 2, 1954, SPI, GWU, file Technological Capabilities; Oral History OH-216, pt. 1, pp. 13–14.

7. Welzenbach, "Din Land," 22–23; Burrows, *This New Ocean*, 156–58; Peebles, *Corona Project*, 18–19; "The Reminiscences of James R. Killian," pt. 1, pp. 13–14; Taubman, *Secret Empire*, 89.

8. Killian, *Sputniks*, 71.

9. McElheny, *Insisting*, 3; Taubman, *Secret Empire*, 91.

10. Welzenbach, "Din Land," 23; R. Cargill Hall, "Strategic Reconnaissance in the Cold War," *Prologue* 28, no. 2 (1996): 118; Peebles, *Shadow Flights*, 78; Peebles, *Corona Project*, 19–20; McElheny, *Insisting*, 293–95; Taubman, *Secret Empire*, 96–97.

11. Interview with Edwin Land, 1984, in Welzenbach, "Din Land," 23; McElheny, *Insisting*, 294.

12. Peebles, *Shadow Flights*, 79; McElheny, *Insisting*, 293–94; Welzenbach, "Din Land," 23; Taubman, *Secret Empire*, 98–99.

13. "Part V, Intelligence: Our First Defense against Surprise," WHO, OSS, SUB, ALPHA, box 16, file Killian Report—Technological Capabilities Panel (2).

14. Comments on the report to the president by the Technological Capabilities Panel of the Science Advisory Committee, WHO, OSANSA, NSC Series, Policy Papers Subseries, box 16, file NSC 5522 Technological Capabilities Panel, p. S21.

15. TCP report, as quoted in Killian, *Sputniks*, 79–80.

16. TCP report, as quoted in Killian, *Sputniks*, 80.

17. Comments on the report to the president by the Technological Capabilities Panel, S21–S22.

18. Comments on the report to the president by the Technological Capabilities Panel, S21–S22.

19. Comments on the report to the president by the Technological Capabilities Panel," s5; "Annex A: Department of State Reaction to Technological Capabilities Panel Recommendations Nos. 7 and 9," June 3, 1955, WHO, OSANSA, NSC Series, Briefing Notes Subseries, box 17, file [Technological Capabilities Panel of the Science Advisory Committee] (3) [1954–56].

20. Comments on the report to the president by the Technological Capabilities Panel," s23.

21. Many historians have described the development of the U-2. For a short summary of Land and his role, see McElheny, *Insisting*, 294–300. For more details, see Beschloss, *May-Day*, or the first two chapters of Brugioni, *Eyeball to Eyeball*; Welzenbach, "Din Land," 23–28; Taubman, *Secret Empire*, 99–102, 113–89.

22. Taubman, *Secret Empire*, 102–5; Beschloss, *May-Day*, 366, 372, 388, 411–12; phone interview with R. Cargill Hall, official historian, National Reconnaissance Office, October 17, 2002; "The Reminiscences of James R. Killian," pt. 2, pp. 35–36; "The Reminiscences of Andrew J. Goodpaster," April 10, 1982, 39–44.

23. Killian, *Sputniks*, 82–83; Hall, "The Eisenhower Administration," 62; Welzenbach, "Din Land," 28; Bissell, "Origins," 15–16; Peebles, *Corona Project*, 19; Taubman, *Secret Empire*, 107.

24. "The Reminiscences of Andrew J. Goodpaster," April 25, 1967, pt. 1, pp. 14–16, 34–36; "The Reminiscences of Andrew Goodpaster, Ann Whitman, Raymond Saulnier, Elmer Staats, Arthur Burns, and Gordon Gray," 15–17; President's appointments, November 24, 1954, AWF, A. W. Diary Series, box 3, file ACW Diary, November 1954 (1); memorandum of conference with the president, November 24, 1954, AWF, A. W. Diary Series, box 3, file ACW Diary, November 1954 (1).

25. Phone interview with R. Cargill Hall, October 17, 2002.

26. Phone interview with R. Cargill Hall, October 17, 2002; "The Reminiscences of James R. Killian," pt. 1, pp. 20–21, pt. 2, pp. 34–36; "The Reminiscences of Dr. Richard M. Bissell," June 5, 1967, 38–41, Columbia Center for Oral History.

27. Memorandum of conference with the president, November 24, 1954; "The Reminiscences of Andrew J. Goodpaster," June 26, 1975, 79–83, Dwight D. Eisenhower Presidential Library.

28. "The Reminiscences of James R. Killian," pt. 1, pp. 19–22; phone interview with R. Cargill Hall, October 17, 2002.

29. Phone interview with R. Cargill Hall, October 17, 2002; "The Reminiscences of James R. Killian," pt. 1, pp. 21–22, pt. 2, pp. 34–37.

30. Gaddis, *The Long Peace*, 198; Alexander, *Holding the Line*, 96–97; Rostow, *Open Skies*, 10, 26–48; Davies and Harris, RAND's Role, 62–63.

31. Gaddis, *The Long Peace*, 198; Alexander, *Holding the Line*, 96–97; Rostow, *Open Skies*, 10, 26–48; Davies and Harris, RAND's Role, 62–63.

32. Rostow, *Open Skies*, 29–30; Karas, *New High Ground*, 97–98; Stassen and Houts, *Eisenhower*, 321–30; Beschloss, *May-Day*, 98–104.

33. Gaddis, *The Long Peace*, 198–99; Beschloss, *May-Day*, 99–100.

34. Gaddis, *The Long Peace*, 199; Albertson, *Eisenhower*, 83–84.

35. The Open Skies proposal remained a fixed element of the U.S. disarmament program following its proposal. "Brief Statement of the Joint Chiefs of Staff Relative to the Problem of Disarmament," January 25, 1956, WHO, OSANSA, NSC Series, Policy Papers Subseries, box 2, file Memorandum for the Secretary of Defense; Stassen and Houts, *Eisenhower*, 334–36.

36. Statement by President Eisenhower at the Geneva Conference of Heads of Government, "Aerial Inspection and Exchange of Military Blueprints," July 21, 1955, in U.S. Department of State, *Documents on Disarmament*, 486, 487–88. See also Stassen and Houts, *Eisenhower*, 339.

37. "Discussion at the 256th Meeting of the National Security Council," July 28, 1955, AWF, NSC Series, box 7, file 256th Meeting of NSC, July 28, 1955; "Discussion at the 257th Meeting of the National Security Council," August 4, 1955, AWF, NSC Series, box 7, file 257th Meeting of NSC, August 4, 1955; Gaddis, *The Long Peace*, 200; Stassen and Houts, *Eisenhower*, 340–41.

38. Article 12 of the Anti-Ballistic Missile Treaty (1972) and Article 15 of the SALT II treaty also used the term. See Herman, *Intelligence Power*, 158–60.

39. Robert R. Bowie, Policy Planning Staff, Department of State, "Memorandum for Mr. Phleger," March 28, 1955, SPI, GWU, file TCP and Overflight.

40. "Tab C: Recommendations under Which Primary Responsibility Was Assigned to Other Agencies, Subject to Coordination with the Central Intelligence Agency," December 22, 1954, WHO, OSANSA, NSC Series, Briefing Notes Subseries, box 17, file [Technological Capabilities Panel of the Science Advisory Committee] (3) [1954–56].

41. "Statement by the President: Summary of Important Facts in the Development by the United States of an Earth Satellite," October 9, 1957, WHO, OSS, SUB, ALPHA, box 23, file Satellites [October 1957–February 1960] (1); Burrows, *This New Ocean*, 169–72.

42. Phone interview with R. Cargill Hall, October 17, 2002; "Tab E: Memorandum RE U.S. Participation in International Geophysical Year: Artificial Satellite Project," June 3, 1955, and "Letter to Mr. Meeker from Mrs. Fleming, Subject: U.S. Participation in International Geophysical Year: Artificial Satellite Project," April 15, 1955, WHO, OSANSA, NSC Series, Briefing Notes Subseries, box 17, file [Technological Capabilities Panel of the Science Advisory Committee] (3) [1954–56].

43. "Memorandum for the Assistant Secretary of Defense (Research and Development), Subject: Scientific Satellite Program for the Department of Defense," May 4, 1955, WHO, OSANSA, NSC Series, Briefing Notes Subseries, box 7, file Earth Satellites (3) [1955–58].

44. "Memorandum for the Assistant Secretary of Defense (Research and Development), Subject: Scientific Satellite Program for the Department of Defense," May 4, 1955.

45. "Memorandum for the Assistant Secretary of Defense (Research and Development), Subject: Scientific Satellite Program for the Department of Defense," May 4, 1955.

46. "Memorandum for the Assistant Secretary of Defense (Research and Development), Subject: Scientific Satellite Program for the Department of Defense," May 4, 1955.

47. "Memorandum for the Assistant Secretary of Defense (Research and Development), Subject: Scientific Satellite Program for the Department of Defense," May 4, 1955.

48. "Memorandum for the Assistant Secretary of Defense (Research and Development), Subject: Scientific Satellite Program for the Department of Defense," May 4, 1955.

49. "Memorandum for Mr. Robert Murphy, Deputy Under Secretary of State, from Alan T. Waterman, Director, National Science Foundation," March 18, 1955, WHO, OSANSA, NSC Series, Briefing Notes Subseries, box 7, file Earth Satellites (3) [1955–1958].

50. "Memorandum for Dr. Alan T. Waterman, Director, National Science Foundation from Mr. Robert Murphy, Deputy Under Secretary of State," April 27, 1955, WHO, OSANSA, NSC Series, Briefing Notes Subseries, box 7, file Earth Satellites (3) [1955–1958].

51. Letter to Donald A. Quarles, assistant secretary of defense (research and development) from Alan T. Waterman, May 13, 1955, WHO, OSANSA, NSC Series, Briefing Notes Subseries, box 7, file Earth Satellites (3) [1955–1958]; "Memorandum for Mr. Robert Murphy, Deputy Under Secretary of State from Alan T. Waterman, Director, National Science Foundation," March 18, 1955, SPI, GWU, file "Bissell and Right of Overflight"; Peebles, *Corona Project*, 22–23; Burrows, *This New Ocean*, 173–74.

52. NSC 5520, "Draft Statement of Policy on U.S. Scientific Satellite Program," May 20, 1955, in Logsdon et al., *Exploring the Unknown*, 1:308; "Memorandum for General Cutler, Subject: U.S. Earth Satellite Program (NSC 5520)," January 18, 1957, WHO, OSANSA, NSC Series, Briefing Notes Subseries, box 7, file Earth Satellites (3) [1955–58]; phone interview with R. Cargill Hall, October 17, 2002.

53. NSC 5520, in Logsdon et al., *Exploring the Unknown*, 1:309; "NSC 5520," SPI, GWU, file NSC 5520, 1.

54. "Tab C: Recommendations," December 22, 1954; phone interview with R. Cargill Hall, October 17, 2002.

55. "Memorandum for General Cutler, Subject: U.S. Earth Satellite Program (NSC 5520)," January 18, 1957, WHO, OSANSA, NSC Series, Briefing Notes Subseries, box 7, file Earth Satellites (3) [1955–58]; NSC 5520, in Logsdon et al., *Exploring the Unknown*, 1:309–24; "Draft Statement of Policy on U.S. Scientific Satellite Program," in *FRUS 1955–1957: United Nations*, 724–30; "NSC 5520," SPI, GWU, file NSC: 5520, 4–5.

56. "Recommendation on Which CIA Has Secondary Responsibility for Reporting to the NSC, Subject to Coordination with Other Agencies," May 12, 1953 [RECOMMENDATION for TCP], WHO, OSANSA, NSC Series, Briefing Note Subseries, box 7, file Earth Satellites (2) [1955–1958], 2, 3; "Discussion at the 250th Meeting of

the National Security Council," May 26, 1955, AWF, NSC Series, box 6, file 250th Meeting of NSC, May 26, 1955.

57. "Memorandum for Mr. James S. Lay Jr. from Nelson A. Rockefeller," May 17, 1955, AWF, Admin. Series, box 31, file Rockefeller, Nelson 1952–55 (5).

58. "Memorandum of the Discussion at the 250th Meeting of the National Security Council, Thursday," May 26, 1955, AWF, NSC Series, box 6, file 250th Meeting of NSC, May 26, 1955; "Draft Report on NSC 5520, U.S. Scientific Satellite Program," November 9, 1956, WHO, OSANSA, NSC Series, Briefing Notes Subseries, box 7, file Earth Satellites (2) [1955–58]; Burrows, *This New Ocean*, 167; Levine, *Missile*, 53.

59. Editorial note, *FRUS*, 11:734.

60. "Memorandum to the President, Subject: Public Information Program with Respect to Implementation of NSC #5520," July 27, 1955, and "Memorandum for General Cutler," January 18, 1957, WHO, OSANSA, NSC Series, Briefing Notes Subseries, box 7, file Earth Satellites (3) [1955–58].

61. "Satellite Programs in the Department of Defense," October 25, 1957, Harlow Records, box 1, file DoD Report to Senate Preparedness Investigating Subcommittee, Missiles (October 1957) (1).

62. "Memorandum for the President, Subject: Project VANGUARD," April 30, 1957, and "Memorandum for: The Secretary of Defense, The Director, Bureau of the Budget, The Director, National Science Foundation, Subject: U.S. Scientific Satellite Program," May 14, 1957, WHO, OSANSA, NSC Series, Briefing Notes Subseries, box 7, file Earth Satellites (2) [1955–58].

63. "Memorandum of Discussion at the 339th Meeting of the National Security Council," October 11, 1957, AWF, NSC Series, box 9, file 339th Meeting of NSC, October 10, 1957.

64. Bowen, *Threshold of Space*, 10–12; "Memorandum for the Deputy Secretary of Defense from C. C. Furnas," July 10, 1956, "Memorandum for Deputy Secretary of Defense from E. V. Murphree, Special Assistant for Guided Missiles," Subject: Use of the JUPITER Re-entry Test Vehicle as a Satellite, July 5, 1956, and "Memorandum for the Assistant Secretary of Defense (R&D) from Homer J. Stewart, Chairman, Advisory Group on Special Capabilities," Subject: VANGUARD and REDSTONE, June 22, 1956, SPI, GWU, file 1957 Jupiter Satellite Launch.

65. "Discussion at the 283rd Meeting of the National Security Council," May 3, 1956, AWF, NSC Series, box 7, file 283rd Meeting of NSC, May 3, 1955; NSC Action No. 1545, as referred to in "Draft Report on NSC 5520," 2, 4–6.

66. "Memorandum of Discussion at the 310th Meeting of the National Security Council," January 24, 1957, and "Memorandum of Discussion at the 322d Meeting of the National Security Council," May 10, 1957, in *FRUS*, 11:743–53; "U.S. Scientific Satellite Program: Cost Estimates," May 9, 1957, WHO, OSS, SUB, DoD Subseries, box 6, file Missiles and Satellites, Vol. 1 (1) [January 1956–May 1957].

67. "Recommendations in the Report to the President by the Technological Capabilities Panel of the Science Advisory Committee, ODM," October 2, 1956, RG 59, GRDS, box 87, NSC 5522 Memoranda; "Memorandum of Discussion at the 339th

Meeting of the National Security Council," October 11, 1957, AWF, NSC Series, box 9, file 339th Meeting of NSC, October 10, 1957.

68. John L. Gaddis attributes this quote to Eisenhower. It was really said by Donald Quarles, quoted in Gaddis, *The Long Peace*, 199. See also Hall, "The Eisenhower Administration," 64; memorandum of discussion at the 339th meeting of the National Security Council, October 11, 1957, AWF, NSC Series, box 9, file 339th Meeting of NSC, October 10, 1957.

69. For a more detailed discussion of space law, its evolution, and its current state, see Benko et al., *Space Law*; Fawcett, *International Law*; Jasentuliyama, *International Space Law*; Gal, *Space Law*; Morenoff, *World Peace*.

70. NSC 5814, June 20, 1958, RG 59, GRDS, box 96, file NSC 5814 (Memorandum), 3–4.

71. Letter to Dr. L. V. Berkner, president, Associated Universities, Inc., from O. G. Villard Jr., National Academy of Sciences, National Research Council, Space Science Board, January 22, 1959, WHO, OSAST (Killian/Kistiakowsky), box 15, file Space [January–June 1959] (6); "Memorandum to J. R. Killian Jr. from David Z. Beckler, Subject: Legal Aspects of Outer Space," April 24, 1959, WHO, OSAST (Killian/ Kistiakowsky), box 15, file Space [January–June 1959] (6); NSC 5814, RG 59, GRDS, 7–8.

72. Taubman, *Secret Empire*, 193.

4. Origins

Dwayne A. Day, "Invitation to Struggle: The History of Civilian-Military Relations in Space," in Logsdon et al., *Exploring the Unknown*, 2:238; portions of this quote paraphrased from Bulkeley, *Sputnik Crisis*, 83.

Robert L. Perry, "A History of Satellite Reconnaissance, Vol. 5: Management of the National Reconnaissance Program, 1960–1965," unpublished official history, National Reconnaissance Office, 1969, 2, declassified November 26, 1997.

1. McDougall, . . . *the Heavens and the Earth*, 89; Davies and Harris, RAND's *Role*, 3–6; Kaplan, *Wizards*, 1–5, 51–63; Burrows, *This New Ocean*, 129.

2. Department of the Navy, Naval Historical Center, "United States Naval Aviation 1910–1995: Part 3. The Twenties 1920–1939," 50, http://www.history.navy.mil /research/histories/naval-aviation-history/united-states-naval-aviation-1910-1995 /part-3-the-twenties-1920-1920.html; Ludwig, *Opening Space Research*; Gruntman, *Blazing the Trail*, 187.

3. Hall, "Earth Satellites: A First Look by the United States Navy," in Hall, *Essays*, 2:253–54; Burrows, *This New Ocean*, 124–25; Robert L. Perry, "Origins of the USAF Space Program, 1945–1956," 1961, SPI, GWU, file Origins of the USAF Space Program, 8–9; Hall, "Earth Satellites," 254–59.

4. Hall, "Earth Satellites," 259, 273–74 (note 41).

5. As quoted in Perry, "Origins of the USAF Space Program," 9.

6. Hall maintains that LeMay rejected the joint program in mid-March 1946 and reconfirmed this stance on April 8, 1946, before the RAND study was com-

missioned. See Hall, "Earth Satellites," 258–60; Perry, "Origins of the USAF Space Program," 9; Augenstein, "Evolution," 3.

7. Burrows, *This New Ocean*, 111–12; author's correspondence with Jim Eckles, Public Affairs Office, White Sands Missile Range, February 27, 2002.

8. Burrows, *This New Ocean*, 125–26; McDougall, . . . *the Heavens and the Earth*, 85–87; Perry, "Origins of the USAF Space Program," 10.

9. McDougall, . . . *the Heavens and the Earth*, 89; Davies and Harris, RAND's *Role*, 3–6; Kaplan, *Wizards*, 1–5, 51–63; Burrows, *This New Ocean*, 129.

10. Burrows, *This New Ocean*, 125–29; Peebles, *Corona Project*, 5–9; Perry, "Origins of the USAF Space Program," 10–11.

11. Day, "Invitation to Struggle," 236; Peebles, *Corona Project*, 5–6; Perry, "Origins of the USAF Space Program," 11–15.

12. Logsdon et al., *Exploring the Unknown*, vol. 1: document II-2, 236–41.

13. Logsdon et al., *Exploring the Unknown*, vol. 1: document II-2, 242; Perry, "Origins of the USAF Space Program," 11–15.

14. Logsdon et al., *Exploring the Unknown*, vol. 1: document II-2, 242.

15. Perry, "Origins of the USAF Space Program," 16; Hall, "Earth Satellites," 258–59.

16. The JRDB was created in June 1946, under the War Department. With Vannevar Bush as chair, it expanded and formalized many of the Aeronautical Board's functions. Its primary responsibility was to prepare an integrated program of research and development to allow the military air arms to evaluate their own programs. It eliminated duplication of effort and had an influence on weapon design. It gave way in 1947 to the Research and Development Board, which emerged in the reorganization of the armed forces in 1947. See Hall, "Earth Satellites," 273–74 (note 41).

17. Perry argues that bureaucratic issues prevented a decision. See Perry, "Origins of the USAF Space Program," 16–17; Hall, "Earth Satellites," 263.

18. Burrows, *This New Ocean*, 126–27.

19. Perry, "Origins of the USAF Space Program," 16–17; Hall, "Strategic Reconnaissance," 107; Hall, "Earth Satellites," 264.

20. Perry, "Origins of the USAF Space Program," 18–20. See also R. Cargill Hall, "Origins of U.S. Space Policy: Eisenhower, Open Skies, and Freedom of Space," in Logsdon et al., *Exploring the Unknown*, 1:214–15; Hall, "Earth Satellites," 263 (notes 57, 275), 264.

21. Perry, "Origins of the USAF Space Program," 20–21; Day, "Invitation to Struggle," 236; Peebles, *Corona Project*, 5–6; Richelson, *America's Secret Eyes in Space*, 3–4.

22. Perry, "Origins of the USAF Space Program," 21–22; Day, "Invitation to Struggle," 236; Augenstein, "Evolution," 4.

23. Perry, "Origins of the USAF Space Program," 22–23; Day, "Invitation to Struggle," 236–37; memorandum for the vice chief of staff, from Lt. Gen. H. A. Craig, air force deputy chief of staff for materiel, "Subject: Earth Satellite Vehicle," January 12, 1948, SPI, GWU, file Air Force in Charge of Space; Augenstein, "Evolution," 4–5.

24. "Statement of Policy for a Satellite Vehicle," Gen. Hoyt S. Vandenberg, vice chief of staff, USAF, January 15, 1948, SPI, GWU, file Air Force in Charge of Space.

25. Perry, "Origins of the USAF Space Program," 21–24; "Statement of Policy"; letter to commanding general, Air Material Command, from Maj. Gen. L. C. Cragie, "Subject: Satellite Vehicle," January 16, 1948, SPI, GWU, file Air Force in Charge of Space; Day, "Invitation to Struggle," 236–37.

26. Merton E. Davies was a mathematician and engineer who in 1947 joined RAND, where he worked on some of its most important studies on space. Amrom H. Katz was a photo interpreter and expert on shutters and lenses who brought fifteen years of experience from the Wright-Patterson AFB Aerial Reconnaissance lab to RAND in 1954. Davies and Harris, RAND's Role, vii; Richelson, America's Secret in Space, 14–15; Peebles, Corona Project, 28.

27. Kecskameti used the earlier RAND research as the technical foundation for a satellite. Predicting a device orbiting at a 350-mile altitude with an orbital period of ninety minutes, he sought an oblique orbit (an orbital path that remains fixed between two defined latitudes) for reconnaissance and outlined requirements for a television reconnaissance system. Paul Kecskameti, "The Satellite Rocket Vehicle: Political and Psychological Problems," RAND Research Memorandum RM-567, October 4, SSB/GCSWS, 1–4.

28. Kecskameti, "The Satellite Rocket Vehicle," 5–8.

29. Kecskameti, "The Satellite Rocket Vehicle," 8–11, 13–14.

30. Kecskameti, "The Satellite Rocket Vehicle," 15–17, 8, 12.

31. The focal length is the distance from the center of a lens to the point where the light rays that make up a visual image come into focus. Thus for a lens with a focal length of 240 inches, the image that the camera sees is focused on a point 240 inches behind the lens. The greater the length, the closer objects appear and the greater the clarity. See Giancoli, Physics, 533–34.

32. Burrows, Deep Black, 26–51; R. Cargill Hall, "Postwar Strategic Reconnaissance and the Genesis of CORONA," in Day et al., Eye in the Sky, 86.

33. Hall, "Strategic Reconnaissance," 107; Peebles, Shadow Flights, 4.

34. Hall, "Postwar Strategic Reconnaissance," 87–88; Hall, "Strategic Reconnaissance," 108.

35. Hall, "Strategic Reconnaissance," 109; Peebles, Shadow Flights, 4–7; Hall, "Postwar Strategic Reconnaissance," 88–90; Taubman, Secret Empire, 35–38; United States Strategic Bombing Survey, as quoted in Hall, "Strategic Reconnaissance," 109. See also Peebles, Shadow Flights, 4–7.

36. Hall, "Strategic Reconnaissance," 109; Peebles, Shadow Flights, 4–7; Hall, "Postwar Strategic Reconnaissance," 88–90; Peebles, Corona Project, 1–3.

37. Hall, "Strategic Reconnaissance," 109–10; Peebles, Shadow Flights, 6–7; Hall, "Postwar Strategic Reconnaissance," 90–92; Peebles, Corona Project, 1–3; Taubman, Secret Empire, 38–39.

38. Layton, And I Was There, 516–17; Day et al., Eye in the Sky, 2–3, 29; McDougall, . . . the Heavens and the Earth, 82–83, 114–15.

39. Hall, "Strategic Reconnaissance," 109–10; Peebles, *Shadow Flights*, 6–7; Hall, "Postwar Strategic Reconnaissance," 90–92.

40. Richelson, *American Espionage*, 100–122; Hall, "Postwar Strategic Reconnaissance," 93–94; Peebles, *Shadow Flights*, 8–9.

41. Richelson, *American Espionage*, 102–27; Hall, "Postwar Strategic Reconnaissance," 93–96; Peebles, *Guardians*, 6–10; Beschloss, *May-Day*, 76–77; Hall, "Strategic Reconnaissance," 111–12.

42. For more information on these programs, see Peebles, *Moby Dick Project*; Peebles, *Shadow Flights*; Johnson and Smith, *Kelly*; Day et al., *Eye in the Sky*, 98, 103–4, 267n38.

43. The RAND studies that followed the original 1946 work on satellite feasibility cover many topics. Most deal with highly technical aspects of satellites, including temperature and pressure densities, flight mechanics, and the identifying of ground launch points. For that reason I do not discuss them here. For more information, see Davies and Harris, *RAND's Role*, 9–15; Perry, "Origins of the USAF Space Program," 25–28; Day, "Invitation to Struggle," 237; Peebles, *Guardians*, 44–45.

44. Memorandum to assistant for evaluation, DCS/D, "Attn: Colonel B. A. Schriever, From Major General C. P. Cabell, Director of Intelligence, Subject: Research and Development on Proposed RAND Satellite Reconnaissance Vehicle," March 17, 1951, SPI, GWU, file Satellite Reconnaissance Proposal 1951.

45. J. E. Lipp, R. M. Salter Jr., R. S. Wehner, R. R. Carhart, C. R. Culp, S. L. Gendler, W. J. Howard, and J. S. Thompson, "Utility of a Satellite Vehicle for Reconnaissance," Project RAND Report R-217, April 1951, SPI, GWU, file Project Rand Report "Utility of a Satellite Vehicle," ix. See also Coolbaugh, "Genesis," 283–84. It should be noted that Robert (Bob) Salter went on to also coauthor the 1954 Project FEEDBACK report before joining the Lockheed Corporation's Missile and Space Division, where he helped write their proposal for the WS-117L satellite under the PIED PIPER proposal process.

46. Lipp et al., "Utility of a Satellite Vehicle," 7–11.

47. Lipp et al., "Utility of a Satellite Vehicle," 1–7, 17–18.

48. Lipp et al., "Utility of a Satellite Vehicle," 40–45.

49. Lipp et al., "Utility of a Satellite Vehicle," 46–62. The report mentions solar energy, although the authors seem not to have considered it; the subject definitely received no detailed discussion.

50. Lipp et al., "Utility of a Satellite Vehicle," 12.

51. Lipp et al., "Utility of a Satellite Vehicle," 12.

52. The report defines resolution with respect to the smallest visible element in a photograph. Thus a resolution of fifty feet would indicate that the smallest item discernible in a photograph as a dot is approximately fifty feet in diameter. Though not as precise as a definition using optical criteria or television lines per millimeter, it is more functional. Variables such as brightness, scene contrast, and exposure time affect resolution. See Lipp et al., "Utility of a Satellite Vehicle," 17–18.

53. Lipp et al., "Utility of a Satellite Vehicle," 12–16, 39; Richelson, *American Espionage*, 174; Perry, "Origins of the USAF Space Program," 28–30; Augenstein, "Evolution," 5–6; Davies and Harris, *RAND's Role*, 25–28.

54. Lipp et al., "Utility of a Satellite Vehicle," 63–69.

55. Memorandum for the deputy chief of staff, development, from Lt. Gen. Thomas D. White, December 18, 1952, "Subject: Satellite Vehicles," SSB/GCSWS, Attachment pp. 2–3.

56. Davies and Harris, *RAND's Role*, 25–29; Richelson, *American Espionage*, 7–9; Peebles, *Corona Project*, 9–10; Taubman, *Secret Empire*, 65–66.

57. Davies and Harris, *RAND's Role*, 25–29; Richelson, *American Espionage*, 7–9; Peebles, *Corona Project*, 9–10, 28; Taubman, *Secret Empire*, 65–67.

58. Perry, "Origins of the USAF Space Program," 30; Augenstein, "Evolution," 5–6.

59. Davies and Harris, *RAND's Role*, 33–34; Hall, "Strategic Reconnaissance," 115; York, *Arms*, 204; Peebles, *Corona Project*, 11–12; Taubman, *Secret Empire*, 51–54.

60. Richard Leghorn also became instrumental in the formulation of the "Open Skies" concept as a consultant to Eisenhower's assistant for disarmament affairs (correspondence with Rick W. Sturdevant, deputy director of history, Air Force Space Command). Peebles, *Corona Project*, 11–12; Davies and Harris, *RAND's Role*, 35.

61. Davies and Harris, *RAND's Role*, 33–34.

62. Peebles, *Corona Project*, 11; Davies and Harris, *RAND's Role*, 35.

63. See Hall, "Strategic Reconnaissance," 115; York, *Arms*, 204–5; Welzenbach, "Din Land," 22; Taubman, *Secret Empire*, 54–56.

64. Peebles, *Shadow Flights*, 63–64.

65. Hall, "Strategic Reconnaissance," 115–16; Peebles, *Corona Project*, 10–12; Davies and Harris, *RAND's Role*, 35–42; Hall, "Origins of U.S. Space Policy," 217.

66. Hall, "Strategic Reconnaissance," 115–16; Peebles, *Corona Project*, 10–12; Davies and Harris, *RAND's Role*, 35–42; Hall, "Origins of U.S. Space Policy," 217–18.

67. Hall, "Strategic Reconnaissance," 116; Peebles, *Corona Project*, 12; Davies and Harris, *RAND's Role*, 44–45; Hall, "Postwar Strategic Reconnaissance," 99; Lewis, *Spy Capitalism*.

68. Peebles, *Corona Project*, 10; Davies and Harris, *RAND's Role*, 44; Augenstein, "Evolution," 5–6; Richelson, *America's Secret Eyes in Space*, 8–9; Perry, "Origins of the USAF Space Program," 30; Hall, "Origins of U.S. Space Policy," 218–19; Coolbaugh, "Genesis," 283–84.

69. Peebles, *Corona Project*, 13; Davies and Harris, *RAND's Role*, 44.

70. Memorandum for the deputy chief of staff, development, December 18, 1952, "Subject: Satellite Vehicles," SSB/GCSWS, Attachment pp. 1–5.

71. Perry, "Origins of the USAF Space Program," 30–31; Hall, "Origins of U.S. Space Policy," 218; Burrows, *This New Ocean*, 174–75; Davies and Harris, *RAND's Role*, 44; Richelson, *America's Secret Eyes in Space*, 8–9; Augenstein, "Evolution," 5–6.

72. Perry, "Origins of the USAF Space Program," 31–32; Davies and Harris, *RAND's Role*, 47; Augenstein, "Evolution," 5–6; Peebles, *Corona Project*, 13–15; Hall,

"Origins of U.S. Space Policy," 218; Robert L. Perry, "A History of Satellite Reconnaissance, Vol. 1: CORONA," unpublished history, National Reconnaissance Office, 1964, revised 1973, declassified November 26, 1997, 2.

73. Perry, "Origins of the USAF Space Program," 32–33; Davies and Harris, *RAND's Role*, 47–48; Richelson, *America's Secret Eyes in Space*, 8–9; Peebles, *Corona Project*, 13–15; Perry, "A History of Satellite Reconnaissance, Vol. 1," 3.

74. For how the current historiography treats the FEEDBACK report, see Richelson, *America's Secret Eyes in Space*, 10–11; Burrows, *This New Ocean*, 175; Hall, "Postwar Strategic Reconnaissance," 105–6, 108–9, 121–23; Peebles, *Corona Project*, 15–16; Hall, "Origins of U.S. Space Policy," 218–19.

75. *Project FEEDBACK Summary Report*, vol. 1, RAND Corporation, Report R-262, March 1, 1954, SPI, GWU, file Project FEEDBACK Report, declassified 1995–96, vii, 149–50, 164–66.

76. *FEEDBACK Report*, 12–13, 126–27.

77. *FEEDBACK Report*, 13–15, 124, 127–32.

78. *FEEDBACK Report*, 132–37.

79. *FEEDBACK Report*, 10–11, 110–25.

80. *FEEDBACK Report*, 17–21.

81. Memorandum to assistant for evaluation, DCS/D, "Attn: Colonel B. A. Schriever, From Major General C. P. Cabell, Director of Intelligence, Subject: Research and Development on Proposed RAND Satellite Reconnaissance Vehicle," March 17, 1951, file Satellite Reconnaissance Proposal 1951, SPI, GWU; *FEEDBACK Report*, 5–6.

82. *FEEDBACK Report*, 93–96.

83. *FEEDBACK Report*, 93–96.

84. Augenstein, "Evolution," 2.

5. WS-117L

Durch, *National Interests*, 36.

1. There is no way to know how often Eisenhower had briefings on WS-117L; no written records document the numbers. In all likelihood they were oral briefings and essentially kept him up to date. Probably General Twining, who was responsible for overhead reconnaissance, conducted them. Phone interview with R. Cargill Hall, October 17, 2002.

2. Coolbaugh worked on a large number of projects during his short time at the WADC, including studies on the performance of unguided missiles in Korea, on the feasibility of a tactical ballistic missile, and on tactical reconnaissance (using rockets), as well as investigation of decoy drones. James S. Coolbaugh, "The Beginning of the Air Force Satellite Program, 1953," unpublished treatise at the request of R. Cargill Hall, USAF historian, summer 1995, SPI, GWU, 1–9 (hereafter Coolbaugh memoirs).

3. Perry, "Origins of the USAF Space Program," 33; Richelson, *America's Secret Eyes in Space*, 9.

4. Perry, "Origins of the USAF Space Program," 33.

5. Coolbaugh memoirs, 44. Perry also refers to "Pied Piper" as the nickname for the industry "investigations." However, he does not associate it with Project 1115. See Perry, "Origins of the USAF Space Program," 38.

6. Coolbaugh memoirs, 44.

7. Perry, "Origins of the USAF Space Program," 32–33; Davies and Harris, *RAND's Role*, 47–48; Peebles, *Corona Project*, 13–15; Perry, "A History of Satellite Reconnaissance, Vol. 1," 3.

8. Peebles, *Corona Project*, 15, 24.

9. Coolbaugh, "Genesis," 284–85.

10. Coolbaugh memoirs, 15; Coolbaugh, "Genesis," 286.

11. Coolbaugh memoirs, 16, 18–19.

12. Coolbaugh, "Genesis," 287; Coolbaugh memoirs, 16.

13. Coolbaugh memoirs, 16–17.

14. Coolbaugh memoirs, 17–18.

15. Coolbaugh memoirs, 17–18; Coolbaugh, "Genesis," 283–300; Hall, "Postwar Strategic Reconnaissance," 105.

16. Biggs also provided two exceptional procurement officers for his study contracts—Frank Daigle and Bob Washburn—who managed to accelerate the process further. See Coolbaugh memoirs, 9–10.

17. Coolbaugh memoirs, 19; Coolbaugh, "Genesis," 288.

18. Coolbaugh memoirs, 19–20.

19. Coolbaugh memoirs, 20–21.

20. Taubman places the date for the system requirement as March 16, 1955; however, no documents support his supposition. Coolbaugh memoirs, 22; Perry, "A History of Satellite Reconnaissance, Vol. 1," 3; "A Chronology of Air Force Space Activities," RG 340, ROSAF, box 367, file 132–59, Satellite Program vol. 2, p. 2; Taubman, *Secret Empire*, 200.

21. Coolbaugh memoirs, 22.

22. RAND also had to use up funding for satellite studies, so Coolbaugh's scheme for spending lab funds was applied to the RAND studies. Coolbaugh memoirs, 23–24, 26–27.

23. A watt hour is a unit of energy equal to the power of 1 watt operating for one hour. Thus 10 watt hours would be a unit of energy equal to the power of 10 watts operating for one hour. Coolbaugh memoirs, 23–24, 29.

24. Coolbaugh memoirs, 29–30.

25. Coolbaugh and General Putt often joked about the funding problems. See Coolbaugh memoirs, 30–31.

26. Coolbaugh memoirs, 30–31. It is clear that Quarles had frequent briefings on the program. His daily diary indicated that the Advanced Reconnaissance System was the subject of several meetings and discussions. See entries for August 24,

October 12, 1955, Quarles Papers, box 1, file Daily Diary 8/15/55–8/15/56 (6); entries for November 8 and 29, 1956, Quarles Papers, box 1, file Daily Diary 8/15/56–4/30/57 (3); phone interview with R. Cargill Hall, October 17, 2002.

27. Virtually every source that identifies General Operational Requirement No. 80 refers to the 1955 date for its release. The only exception that is evident and worthy of note is a document titled "The CORONA Story," which does not provide a date beyond the fact that General Operational Requirement No. 80 was released in 1954. Why this account differs is unclear. See Oder et al., "The CORONA Story," 4.

28. "General Operational Requirement for a Reconnaissance Satellite Weapon System," GOR No. 80, March 15, 1955, revised September 26, 1958, SPI, GWU, file General Operational Requirement; Kenneth E. Greer, "CORONA," Studies in Intelligence, Supplement 17 (Spring 1973): 3, reprinted in Ruffner, CORONA, 3–24.

29. Hall, "Postwar Strategic Reconnaissance," 106.

30. Coolbaugh memoirs, 33–34; Coolbaugh, "Genesis," 292–93; Peebles, Corona Project, 25–26; phone interview with R. Cargill Hall, October 17, 2002.

31. Coolbaugh memoirs, 36; Perry, "A History of Satellite Reconnaissance, Vol. 1," 3.

32. Coolbaugh memoirs, 36; "Truax Memoirs," unpublished manuscript chapters, SPI, GWU, file Truax Memoirs, 299.

33. Perry, "Origins of the USAF Space Program," 39–40; "Truax Memoirs," 299; Perry, "A History of Satellite Reconnaissance, Vol. 1," 3; Peebles, Corona Project, 24; Coolbaugh, "Genesis," 294–95.

34. Perry, "Origins of the USAF Space Program," 39–40; "Truax Memoirs," 299; Perry, "A History of Satellite Reconnaissance, Vol. 1," 3; Peebles, Corona Project, 24; Coolbaugh, "Genesis," 294–95.

35. Perry, "Origins of the USAF Space Program," 38; Peebles, Corona Project, 24; Richelson, America's Secret Eyes in Space, 13.

36. Captain Truax was head of the program office only briefly. In August 1956 air force colonel Frederick C. E. (Fritz) Oder became head of the Air Force Reconnaissance Office. Truax remained part of the program working under Oder. See Day et al., Eye in the Sky, 107, 109.

37. Peebles, Corona Project, 24; Coolbaugh memoirs, 43, 46; Coolbaugh, "Genesis," 295.

38. This effort slowed somewhat because of attempts to use component parts of WS-117L as part of the scientific satellite proposal. This effort eventually halted when the administration separated civilian and military programs. See Perry, "Origins of the USAF Space Program," 41–52.

39. Perry, "Origins of the USAF Space Program," 52–53; Perry, "A History of Satellite Reconnaissance, Vol. 1," 3–4; Richelson, America's Secret Eyes in Space, 13; "A Chronology of Air Force Space Activities," 2:2.

40. Peebles, Corona Project, 25. For a more detailed discussion of the second stage, see "Weapon System 117L Preliminary Development Plan, Advanced

Reconnaissance System," January 14, 1956, RG 340, ROSAF, box 118, file 190–56, "Scientific Earth Satellite," 6–10; Perry, "A History of Satellite Reconnaissance, Vol. 1," 4.

41. Perry, "Origins of the USAF Space Program," 54; Coolbaugh, "Genesis," 297.

42. Letter to Eisenhower from Neil McElroy, May 7, 1958, AWF, DDE Diary Series, box 33, file Toner Notes—May 1958 (2); Perry, "Origins of the USAF Space Program," 54–55; Perry, "A History of Satellite Reconnaissance, Vol. 1," 3–4. As an indication of how complicated and confused accounts of the early program can be, Anthony Kenden, Curtis Peebles, and Philip J. Klass give a contract date with Lockheed of June 30, 1956. This may have been the original decision, but signing had to wait for October for funding reasons. Kenden, "U.S. Reconnaissance Satellite Programmes," 243; Klass, Secret Sentries, 82–83; Peebles, Corona Project, 25.

43. Coolbaugh memoirs, 47–48.

44. "Truax Memoirs," 301–2.

45. "Truax Memoirs," 303; Perry, "A History of Satellite Reconnaissance, Vol. 1," 3–5; Hall, "The Eisenhower Administration," 64; Oder et al., "The CORONA Story," 5.

46. Perry, "A History of Satellite Reconnaissance, Vol. 1," 5–6. There is only one account of the program that implies that ample funding was available. This is clearly an error; I include it only to indicate the level of confusion about the program. See Hochman and Wong, Satellite Spies, 100–101.

47. Donald Quarles died in 1959. Although his personal correspondence, lists of meetings, and speeches are at the Eisenhower Library, nothing relating to his decisions on satellite reconnaissance appears in this collection. See Quarles Papers, boxes 1–27; Perry, "A History of Satellite Reconnaissance, Vol. 1," 6–7; Hall, "Origins of U.S. Space Policy," 222.

48. Perry, "Origins of the USAF Space Program," 41–52.

49. "Satellite Programs in the Department of Defense," October 25, 1957, Harlow Records, box 1, file DoD Report to Senate Preparedness Investigating Subcommittee, Missiles (October 1957) (1); "Discussion at the 339th Meeting of the National Security Council, 10 October 1957," AWF, NSC Series, box 9, file 339th Meeting of NSC, October 10, 1957; Perry, "A History of Satellite Reconnaissance, Vol. 1," 6–14.

50. Perry, "A History of Satellite Reconnaissance, Vol. 1," 14–17.

51. Oder and Schriever were certainly not the only ones viewing the "space for peace" policy in a negative light. Richard Leghorn put forward a memorandum in July 1957 concerning the role of space for peace and satellite reconnaissance. His proposal called for the direct linkage of satellite reconnaissance and arms-control measures. This offered a political windfall for the United States and would allow for the satellite reconnaissance effort to come out of the shadows when the time was ripe. There is no evidence that this information was available to General Schriever or Oder. It is unlikely that this memorandum had a major impact on the administration. Eisenhower had already linked the idea of overhead reconnaissance and arms control in his Open Skies proposal, and there is no indication that the mem-

orandum reached the key players in his administration. "The Reconnaissance Satellite and 'Space-for-Peace' Political Action," July 31, 1957, WHO, OSAST, box 3, file Space-Satellites [July 1956–February 1960]; "Richard S. Leghorn (PBCFIA) to D. Z. Beckler, re: Political Action and Unauthorized Overflight of the USSR," July 26, 1956, and "Richard Leghorn to Mr. Beckler (PSAC) re: The Reconnaissance Satellite and 'Space for Peace' Political Action," July 31, 1957, WHO, OSAST, box 3, file Space-Satellites [July 1956–February 1960].

52. The two accounts of Perry and Oder are remarkably similar. See Perry, "A History of Satellite Reconnaissance, Vol. 1," 17–18; Oder et al., "The CORONA Story," 10–13.

53. Perry, "A History of Satellite Reconnaissance, Vol. 1," 18–22; Peebles, *Guardians*, 45.

54. Perry, "Origins of the USAF Space Program," 41–56; Perry, "A History of Satellite Reconnaissance, Vol. 1," 22–23.

55. The Western Development Division officially became the Ballistic Missile Division in August 1957.

56. Perry, "The History of Satellite Reconnaissance, Vol. 1," 24.

57. Hall, "Postwar Strategic Reconnaissance," 107–8; Richelson, *America's Secret Eyes in Space*, 19; Peebles, *Corona Project*, 30; Klass, *Secret Sentries*, 86; Kenden, "U.S. Reconnaissance Satellite Programmes," 244.

58. William Burrows mentions a drop-film system associated with WS-117L— perhaps an addition in 1957, although the date is still very unclear. This variation, which eventually became CORONA, receives further discussion in chapter 6, along with the RAND work involving such a system.

59. Peebles, *Corona Project*, 25.

60. Robert L. Perry, "A History of Satellite Reconnaissance, Vol. 2-A: SAMOS," unpublished history, National Reconnaissance Office, 1963, revised 1973, declassified June 2001, 5.

61. Hall, "Postwar Strategic Reconnaissance," 107–8; Ruffner, CORONA, 3; Peebles, *Corona Project*, 25; Oder et al., "CORONA Story," 4.

62. Divine, *Sputnik Challenge*, 11.

6. Satellite Photography

Kecskameti, "The Satellite Rocket Vehicle," 8.

1. Levine, *Missile*, 58.

2. For a more detailed account of *Sputnik* and the resulting crisis, see Bulkeley, *Sputnik Crisis*; Divine, *Sputnik Challenge*; Burrows, *This New Ocean*; McDougall, . . . *the Heavens and the Earth*.

3. Peebles, *Corona Project*, 39–42; Bottome, *Missile Gap*, 30–45, 87–88, 92–106; Levine, *Missile*, 59, 61–66; Taubman, *Secret Empire*, 213–14; Krug, *Presidential Perspectives*, 23–24.

4. Hethloff, *Suddenly*, 1, 11; Levine, *Missile*, 59–60.

5. McDougall, . . . *the Heavens and the Earth*, 142, 152–53; Licklider, "The Missile Gap Controversy," 615, 609; Krug, *Presidential Perspectives*, 23–24; Bottome, *Missile Gap*, 79–91.

6. "Open Letter to President Eisenhower," *New York Times*, November 7, 1957, WHO, OSS, SUB, ALPHA, box 23, file Satellites [October 1957–February 1960] (1); Ambrose, *Eisenhower: Soldier and President*, 449–50; Alexander, *Holding the Line*, 213–14; Eisenhower, *White House Years*, 2:205–6.

7. The information that the U-2 produced was accurate enough for the CIA to predict a Soviet satellite launch in November 1957. Klass, *Secret Sentries*, 30–38, 51; Killian, *Sputniks*, 6.

8. "Memorandum for Secretary Quarles and General Cutler," October 8, 1957, and "Statement by the President," October 9, 1957, WHO, OSS, SUB, ALPHA, box 23, file Satellites [October 1957–February 1960] (1); "Memorandum of Conference with the President, 8 October 1957," AWF, DDE Diary Series, box 27, file October '57 Staff Notes (2); statement by the president, "Summary of Important Facts in the Development by the United States of an Earth Satellite," October 9, 1957, AWF, Admin. Series, box 37, file U.S. Satellites.

9. Some personnel in the administration thought such an approach not the best. See "Memorandum for Mr. Victor Cooley from Mr. David Z. Beckler, Subject: Satellites and Missiles, 8 October 1957," WHO, OSAST, box 3, file Space.

10. Ambrose, *Eisenhower: The President*, 430. See also Krug, *Presidential Perspectives*, 24–25.

11. "Memorandum of Conference with the President, 16 October 1957," AWF, DDE Diary Series, box 27, file October '57 Staff Notes (2); Killian, *Sputniks*, xv, 11–20, 122–23; McElheny, *Insisting*, 310–17; Killian, *Education*, 326–28.

12. "Memorandum of a Conference, President's Office, White House," October 8, 1957, in *FRUS 1955–1957: United Nations*, 755–56; Eisenhower, *White House Years*, 2:210–19.

13. The flurry of activity included the army's proposal for a reconnaissance satellite that virtually duplicated the WS-117L program. See "Briefing on Army Satellite Program," November 19, 1957, WHO, OSAST (Killian/Kistiakowsky), box 15, file Space [November 1957] (2); letter to Neil McElroy, secretary of defense, from Eisenhower, October 17, 1957, WHO, OSS, SUB, DoD Subseries, box 11, file Secretary of Defense.

14. "Defense Space Projects Supplement to Department of Defense Report to National Security Council on Status of United States Military Programs as of 30 June 1959: Annex B," October 26, 1959, WHO, OSANSA, NSC Series, Status of Projects Subseries, box 8, file NSC 5912 (2); letter to Eisenhower from Secretary of Defense Neil McElroy, January 29, 1958, WHO, OSS, SUB, DoD Subseries, box 8, file Missiles and Satellites, Military Satellite Program, December 31, 1958.

15. Hall, "Postwar Strategic Reconnaissance," 108–9. See also Davies and Harris, *RAND's Role*, 69–70; memorandum to J. L. Hult from A. H. Katz, "Subject: Recom-

mendations, Formal and Informal, on Reconnaissance Satellite Matters," RAND Corporation Memo No. M-3008, 5–11–1959, SPI, GWU, file "Amrom Katz Recommendations on Reconnaissance Satellites."

16. Davies and Harris, RAND's Role, 69–70; Hall, "Postwar Strategic Reconnaissance," 108–9; Peebles, Corona Project, 27–28.

17. J. H. Huntzicker and H. A. Lieske, "Physical Recovery of Satellite Payloads—A Preliminary Investigation," RAND Corporation Research Memorandum, RM-1811, June 26, 1956, SSB/GCSWS, 1; Peebles, Corona Project, 27–28; Taubman, Secret Empire, 225–62.

18. Huntzicker and Lieske, "Physical Recovery," 1–7.

19. Huntzicker and Lieske, "Physical Recovery," 8–11.

20. Huntzicker and Lieske, "Physical Recovery," 12–15.

21. Memorandum to J. L. Hult from A. H. Katz, "Subject: Recommendations, Formal and Informal, on Reconnaissance Satellite Matters."

22. The problems relating to heating of the capsule during reentry somewhat eased with the discovery of the ablative heat shield. During a summer study in 1956, the ablative nose cone being developed for ICBMs emerged as a solution. See Richelson, America's Secret Eyes in Space, 14–16.

23. Richelson, America's Secret Eyes in Space, 14–15; Peebles, Corona Project, 27–28; Davies and Harris, RAND's Role, vii.

24. By September 1957, in the face of severe financial restrictions, the new Itek (short for Information Technology) corporation took over the Boston Lab's role. Richard Leghorn founded Itek to continue work on developing high-altitude cameras. See Peebles, Corona Project, 28–30.

25. Other people had already pointed out the value of an IRBM as a booster for a multistage rocket. Ramo-Woolridge had proposed such a concept on April 1, 1957. See Ramo-Woolridge Document GM67.3–49, Subject: "Proposed Use of IRBM as Booster for Multi-stage Vehicles, 1 April 1957," in S. A. Grassly, "Documentary History of DISCOVERER," November 1971, History Office, Chief of Staff, Space & Missiles Systems Organization, SPI, GWU, Document No. 3.

26. "Letter to DDE from James Killian," December 20, 1956, WHO, OSANSA, NSC Series, Subject Subseries, box 7, file President's Board of Consultants on Foreign Intelligence Activities, First Report to the President [December 1956–August 1958] (1).

27. "Memorandum of Conference with the President," October 24, 1957, AWF, DDE Diary Series, box 27, file October '57 Staff Notes (1); "Memorandum of Conference with the President," October 28, 1957, WHO, OSS, SUB, ALPHA, box 6, file Board of Consultants [on Foreign Intelligence Activities] (4).

28. McElheny, Insisting, 316–18; Peebles, Corona Project, 42–43; Albert D. Wheelon, "CORONA: A Triumph of American Technology," in Day, Eye in the Sky, 32–33; Hall, "Postwar Strategic Reconnaissance,"109–10.

29. Peebles, Corona Project, 42–43; Oder et al., "The CORONA Story," 14–15; Wheelon, "CORONA," 32–33; Hall, "Postwar Strategic Reconnaissance," 109–10.

30. The history by Oder and that by Perry and Greer are virtually identical in wording in many respects. Oder et al., "The CORONA Story," 13–14; Perry, "A History of Satellite Reconnaissance, Vol. 1," 26; Greer, "CORONA," 4–5.

31. Perry, "A History of Satellite Reconnaissance, Vol. 1," 24–27.

32. Davies and Harris, RAND's Role, 95; Peebles, Guardians, 46.

33. "Priorities for Ballistic Missiles and Satellite Programs," January 22, 1958, AWF, NSC Series, box 9, file 352nd Meeting of NSC, January 22, 1958; "Memorandum for the Special Assistant to the President for National Security Affairs, Subject: Priorities for Ballistic Missile and Space Programs," August 10, 1959, WHO, OSANSA, NSC Series, Briefing Notes Subseries, box 14, file [Outer Space Policy and Activities] (2) [1958–60].

34. Richelson, America's Secret Eyes in Space, 29–30; Davies and Harris, RAND's Role, 95–97; "Priorities for Ballistic Missiles and Space Programs," May 13, 1959, AWF, NSC Series, box 11, file 406th Meeting of NSC, May 13, 1959; "Priorities for Ballistic Missiles and Space Programs," August 18, 1959, AWF, NSC Series, box 11, file 417th Meeting of NSC, August 18, 1959; "Memorandum for the Special Assistant to the President for National Security Affairs, from James H. Douglas, Subj: Priorities for Satellite Programs," WHO, OSANSA, NSC Series, Briefing Notes Subseries, box 14, file [Outer Space Policy and Activities] (1) [1958–60]; "Memorandum for the Special Assistant to the President for National Security Affairs, from Donald A. Quarles, Subj: NSC 1846," April 21, 1958, WHO, OSS, SUB, DoD Subseries, box 8, file Missiles and Satellites, Military Reconnaissance Satellite Program; "Memorandum for the Secretary of Defense, Subj: NSC 1846 'Priorities for Ballistic Missiles and Satellite Programs,'" April 21, 1958, WHO, OSS, SUB, DoD Subseries, box 8, file Missiles and Satellites, Military Reconnaissance Satellite Program; "Letter to the President, from Secretary McElroy," May 7, 1958, WHO, OSS, SUB, DoD Subseries, box 8, file Missiles and Satellites, Military Reconnaissance Satellite Program.

35. M. E. Davies et al., "A Family of Recoverable Reconnaissance Satellites," RAND Corporation Research Memorandum, RM-2012, November 12, 1957, SPI, GWU, file "A Family of Recoverable Reconnaissance Satellites," 1–4, 27–32; Davies and Harris, RAND's Role, 69–72, 76–89; Peebles, Corona Project, 42–44.

36. Davies et al., "A Family of Recoverable Reconnaissance Satellites," 27–32.

37. Davies et al., "A Family of Recoverable Reconnaissance Satellites," iii–v, 5–19, 21–23.

38. Davies et al., "A Family of Recoverable Reconnaissance Satellites," iii–v, 17–19, 21–23.

39. Davies et al., "A Family of Recoverable Reconnaissance Satellites," 23–26.

40. "An Earlier Reconnaissance Satellite System," RAND Corporation Recommendation to the Air Staff, November 12, 1957, SSB/GCSWS, 1–19.

41. Lockheed Aircraft Corporation, Missile Systems Division, "WS 117L Development Plan for Program Acceleration," Contract AF 04(647)-97, LMSD-2832, January 6, 1958, pp. 1–2, SPI, GWU, file "Acceleration of WS-117L"; Oder et al., "The CORONA Story," 17–18.

42. "ws-117L Development Plan for Program Acceleration," 1–2, section 1.1.1, System Considerations; Oder et al., "The CORONA Story," 18.

43. "ws-117L Development Plan for Program Acceleration," section 1.1.1.2, Program I, Section 1.1.1.3, Program II.

44. "ws-117L Development Plan for Program Acceleration," section 1.1.1.3.

45. "ws-117L Development Plan for Program Acceleration," sections 1.1.1.4–1.1.1.6.

46. "ws-117L Development Plan for Program Acceleration," section 1.2.

47. "ws-117L Development Plan for Program Acceleration," section 1.2.

48. Oder et al., "The CORONA Story," 15–16, 20; Peebles, Corona Project, 42–43; Greer, "CORONA," 4–5; McElheny, Insisting, 316–18.

49. Hall, "Postwar Strategic Reconnaissance," 111–12; Oder et al., "The CORONA Story," 18–20; memorandum of conference with the president, February 7, 1958, WHO, OSS, SUB, ALPHA, box 14, file Intelligence Matters (4); memorandum of conference with the president, February 10, 1958, AWF, DDE Diary Series, box 30, file Staff Notes February 1958.

50. Memorandum of conference with the president, February 10, 1958, WHO, OSS, SUB, ALPHA, box 14, file Intelligence Matters (4); memorandum of conference with the president, February 10, AWF, DDE Diary Series, box 30, file Staff Notes February 1958.

51. Memorandum of conference with the president, February 10, 1958, WHO, OSS, SUB, ALPHA; memorandum of conference with the president, February 10, 1958, AWF.

52. The rivalry between the services for funding for space programs, based on very thin justification, serves as one of the clearest examples of the interservice rivalry that so troubled Eisenhower. See Richelson, America's Secret Eyes in Space, 24–27; Bowen, Threshold of Space, 16–28.

53. Memorandum for undersecretary of the air force from Maj. Gen. H. C. Donnelly, assistant deputy chief of staff, plans and programs, "Subject: Proposed Statements for Senate Hearings on Missiles," November 21, 1957, RG 340, ROSAF, box 134, file 21–54 Guided Missiles General; "USAF Pushes Pied Piper Space Vehicle," 26–27; Klass, Secret Sentries, 87–88; Richelson, America's Secret Eyes in Space, 20, 26–27.

54. The lack of radiated signals from the film-return system helped to decrease the risk of Soviet reaction. More difficult to detect, the film-drop camera had the potential to limit the political repercussions and decrease the risks of Soviet countermeasures. Greer, "CORONA," 5–6; Burrows, Deep Black, 86; Hall, "Postwar Strategic Reconnaissance," 111–12; Day, "Strategy for Reconnaissance," in Day et al., Eye in the Sky, 138–39.

55. Hall, "Postwar Strategic Reconnaissance," 112; Burrows, Deep Black, 55–56.

56. Memorandum of conference with the president, January 17, 1958, WHO, OSS, SUB, ALPHA, box 14, file Intelligence Matters (4).

57. The origin of the name CORONA is still somewhat debated. The usual explanation is that it came from the typewriter, a Smith-Corona, being used in March 1958 during a meeting that finally worked out the satellite concept. However, some

old-timers argue it had more to do with the cigar one of the team smoked. No matter the source, it adds flavor to the program. See Peebles, *Corona Project*, 44; Day et al., *Eye in the Sky*, 6; memorandum for the record, April 21, 1958, WHO, OSS, SUB, ALPHA, box 14, file Intelligence Matters (5).

58. The project manager for WS-117L, Fritz Oder, took no umbrage at the creation of CORONA. Intimately familiar with the problems in finding resources for the WS-117L system, he was positive that money for both programs would not be available from the air force. See Richelson, *America's Secret Eyes in Space*, 26.

59. Oder et al., "The CORONA Story," 15–21; Perry, "A History of Satellite Reconnaissance, Vol. 1," 37–46; Hall, "Postwar Strategic Reconnaissance," 113; Richelson, *America's Secret Eyes in Space*, 26–28; Peebles, *Corona Project*, 43–46; Ruffner, CORONA, 5–6; Wheelon, "CORONA," 33–34.

60. Perry, "A History of Satellite Reconnaissance, Vol. 1," 39–43; Richelson, *America's Secret Eyes in Space*, 27–28; Peebles, *Corona Project*, 45; Ruffner, CORONA, 6–7; Oder et al., "The CORONA Story," 21–22.

61. "Proposed Initial Press Release," November 6, 1958, WHO, OSS, SUB, ALPHA, box 12, file Discoverer [March–December 1958] (1); Perry, "A History of Satellite Reconnaissance, Vol. 1," 39–44; Memorandum to Lockheed Aircraft Corporation from commander AFBMD, March 12, 1958, in Grassly, "Documentary History," Document 30; memorandum for Major General Funk, "Subject: Reorientation WS 117L Program IIA," March 14, 1958, in Grassly, "Documentary History," Document 32.

62. Day et al., *Eye in the Sky*, 6–9.

63. Peebles, *Guardians*, 47–49; Peebles, *Corona Project*, 46–49.

64. The best accounts to date of the CORONA program are Day et al., *Eye in the Sky*, and Peebles, *Corona Project*. A small sampling of the documents and imagery appears in Ruffner, CORONA; Oder et al., "The CORONA Story."

7. SENTRY/SAMOS

Dwight D. Eisenhower, October 9, 1957, quoted in Ambrose, *Eisenhower: The President*, 430.

Perry, "A History of Satellite Reconnaissance, Vol. 2-A," 136.

1. McElheny, *Insisting*, 328; "Memorandum of Conference with the President," March 12, 1958, AWF, DDE Diary Series, box 31, file Staff Notes, March 1958 (2); "Defense Space Projects Supplement to Department of Defense Report to National Security Council on Status of United States Military Programs as of 30 June 1959: Annex B."

2. Hall, "The Eisenhower Administration," 65; Augenstein, "Evolution," 11–12.

3. "Memorandum for the President," July 29, 1958, AWF, Admin. Series, box 34, file Stans, Maurice H. (director Bureau of the Budget) 1958 (1); Bowen, *Threshold of Space*, 24–25; Richard J. Barber Associates, Inc., "The Advanced Research Projects Agency, 1957–1964," December 1975, SPI, GWU, file "ARPA History."

4. "Defense Space Projects Supplement to Department of Defense Report to National Security Council on Status of United States Military Programs as of 30 June 1959: Annex B"; memorandum for the secretary of the air force from Roy Johnson, director, Advanced Research Projects Agency, "Subject: Reconnaissance Satellites and Manned Space Exploration," February 28, 1958, SSB/GCSWS; Perry, "A History of Satellite Reconnaissance, Vol. 1," 43–44, 55–56.

5. Barber Associates, Inc., "The Advanced Research Projects Agency," III-11–III-12. For a brief discussion of the program's complexities, see memorandum for General Cutler from Bureau of the Budget, June 20, 1958, WHO, OSANSA, NSC Series, Briefing Notes Subseries, box 14, file [Outer Space Policy and Activities] (2) [1958–60].

6. Memorandum for the special assistant to the president for national security affairs, July 28, 1958, and defense recommendation on the reconnaissance satellite (submitted to the NSC on July 3, 1958), July 31, 1958, WHO, OSANSA, NSC Series, Briefing Notes Subseries, box 7, file Earth Satellites (1) [1955–58]; "Attachment B, Memorandum for Dr. Killian, Subject: Notes on the Space Panel Meeting," July 8, 1958, WHO, OSAST (Killian/Kistiakowsky), box 15, file Space Notebook [Killian, 1958–59] (2); "National Space Activities: A Brief Summary," August 12, 1958, WHO, OSS, SUB, DoD Subseries, box 8, file Missiles and Satellites, National Space Activities—A Brief Summary, August 12, 1958.

7. Memorandum for the space file from R. O. Piland, "Subject: Space Budget and Program Responsibilities," July 14, 1958, WHO, OSAST (Killian/Kistiakowsky), box 15, file Space Notebook [Killian, 1958–59] (3).

8. There are many examples of leaked information. See "SENTRY Satellite Shot Planned for Dec. 15"; "USAF Pushes Pied Piper Space Vehicle"; "Lockheed Builds WS-117L Nerve Center"; "Memorandum for Dr. James R. Killian Jr. from Robert Cutler, Special Assistant to the President, Subject: Scientific and Reconnaissance Satellites," January 20, 1958, WHO, OSAST, box 3, file Space Satellites.

9. Letter from McElroy to Eisenhower, January 29, 1958, WHO, OSS, SUB, DoD Subseries, box 8, file Missiles and Satellites, Military Satellite Program, December 31, 1958; Peebles, Corona Project, 53; Perry, "A History of Satellite Reconnaissance, Vol. 1," 73; Perry, "A History of Satellite Reconnaissance, Vol. 2-A," 25; "Memorandum for Mr. Lay from Gordon Gray," March 3, 1959, WHO, OSANSA, NSC Series, Briefing Notes Subseries, box 7, file [Earth Satellites (1) [1955–58]; "National Space Activities: A Brief Summary," August 12, 1958, WHO, OSS, SUB, DoD Subseries, box 8, file Missiles and Satellites, National Space Activities—A Brief Summary, August 12, 1958.

10. Lee Bowen, "The Threshold of Space: The Air Force in the National Space Program 1945–1959," September 1960, SSB/GCSWS, 31; Letter from McElroy to Eisenhower, January 29, 1958; Peebles, Corona Project, 53; Perry, "A History of Satellite Reconnaissance, Vol. 1," 73; Perry, "A History of Satellite Reconnaissance, Vol. 2-A," 25; "Memorandum for Mr. Lay from Gordon Gray," March 3, 1959; Richard J. Barker Associated, Inc., "The Advanced Research Projects Agency, 1957–1964," III-12.

11. ARPA Order No. 38-59 separated out MIDAS as its own system. See "Memorandum for the Secretary of Defense from Malcolm A. MacIntyre, Subject: Operational Responsibility in Respect to MIDAS," May 8, 1959, RG 340, ROSAF, box 367, file 132-59, Satellite Programs vol. 1; Richard J. Barker Associated, Inc., "The Advanced Research Projects Agency, 1957–1964," III-12.

12. A memorandum of July 29, 1958, places the WS-117L budget from ARPA at $186 million. See "Memorandum for the President," July 29, 1958, AWF, DDE Diary Series, box 35, file Staff Memos July 1958 (1). No accurate information on budget amounts is yet available. Figures for the WS-117L budget for FY 1959 range from $100 million to $239 million. See "Memorandum of Conference with the President," November 11, 1957, AWF, DDE Diary Series, box 28, file November '57 Staff Notes; "Chart on Outer Space Vehicles" (undated), WHO, OSAST (Killian/Kistiakowsky), box 15, file Space [December 1957] (3); Richard J. Barber Associated, Inc., "The Advanced Research Projects Agency, 1957–1964," III-12–III-13.

13. SENTRY Space System Development Plan, January 30, 1959, RG 340, ROSAF, box 367, file 132-59 "Satellite Program Vol. 2," I-1-2–I-1-3. See also "Letter to the President from Secretary of Defense McElroy," May 7, 1958, WHO, OSS, SUB, DoD Subseries, box 8, file Missiles and Satellites, Military Reconnaissance Satellite Program, March 31, 1958.

14. SENTRY Space System Development Plan, I-2-1–I-2-3.

15. All individual subsystems for WS-117L received a letter designation for management and development purposes. The photographic portion was subsystem E; electronic intelligence, subsystem F. See Perry, "A History of Satellite Reconnaissance, Vol. 2-A," 8.

16. Perry, "A History of Satellite Reconnaissance, Vol. 2-A," iii, 8–9, 16, 19; SENTRY Space System Development Plan, I-2-3–I-2-9, I-2-11–I-2-12; Day, "The Development and Improvement of the CORONA Satellite," in Day et al., *Eye in the Sky*, 71–72.

17. Day, "Development and Improvement," 63, 74–75.

18. The recovery program for SAMOS experienced a great deal of opposition and turmoil in 1958–60. See Perry, "A History of Satellite Reconnaissance, Vol. 2-A," 20–27, 74–75, 84–85; "SAMOS," undated, SPI, GWU, 1; Day, "Development and Improvement," 71–72.

19. SENTRY Space System Development Plan, I-2-3, I-2-5, I-2-7, I-2-12.

20. SENTRY Space System Development Plan, I-2-12–I-2-13.

21. SENTRY Space System Development Plan, I-3-2–I-3-3.

22. For the period up to and including FY 1958, the overall costs for SENTRY amounted to approximately $79,340,000. Lockheed received most of these funds for the development of the various subsystems. Additional funding went to facilities ($5,718,000), the Rome Air Development Center ($2,820,000), the Air Force Cambridge Research Center ($2,842,000), and other, smaller-cost elements. See SENTRY Space System Development Plan, I-3-6, I-5-1–I-5-12, II-1-2.

23. Perry, "A History of Satellite Reconnaissance, Vol. 2-A," 10–13; McElheny, *Insisting*, 331–33.

24. Perry, "A History of Satellite Reconnaissance, Vol. 2-A," 13–15; Burrows, *This New Ocean*, 227–28.

25. Perry, "A History of Satellite Reconnaissance, Vol. 2-A," 62–68, 151–53, 166–68, 170–73; "SAMOS," 1–4; letter to Eisenhower from James H. Douglas, deputy director of defense, November 19, 1960, WHO, OSANSA, NSC Series, Briefing Notes Subseries, box 13, file Missiles and Military Space Programs (1) [1955–61].

26. Day, "Development and Improvement," 73–74.

27. Memorandum to the director, Advanced Research Projects Agency, "Subject: MIDAS Progress Report," August 31, 1959, RG 340, ROSAF, box 367, file 132-59 "Satellite Program Vol. 2."

28. "ARPA Order No. 38–59," November 5, 1958, SSB/GCSWS.

29. "Report of the Ad Hoc Technical Advisory Board to the Advanced Research Projects Agency on the Evaluation of Technical Feasibility of Infrared Missile Defense Alarm from Satellite Vehicles," February 26, 1959, RG 340, ROSAF, box 367, file 132-59 "Satellite Program Volume 1"; "Report of the Early Warning Panel," March 13, 1959, WHO, OSAST (Killian/Kistiakowsky), box 4, file "ICBM-Early Warning [November 1957–December 1960]," 10–11.

30. MIDAS Operational System Description, December 18, 1959, RG 340, ROSAF, box 367, file 132-59 "Satellite Program," iii, I-1–I-3.

31. Earlier in the program Lockheed Missile and Space Division predicted twenty satellites would be necessary. See Missile Defense Alarm System Development Plan, January 30, 1959, revised June 5, 1959, RG 340, ROSAF, box 367, file 132-59 "Satellite Programs Vol. 2," I-2–2; MIDAS presentation, August 12, 1959, LMSD-445610, NARA II, RG 340, ROSAF, box 368, file 132-59 "Satellite Program," 1–3.

32. The probable locations of the three stations were New Boston, New Hampshire; Ottumwa, Iowa; and Fort Stevens, Oregon. See MIDAS operational system description, December 18, 1959, RG 340, ROSAF, box 367, file132-59 "Satellite Program," 2–5.

33. For a discussion of the satellite-handling and -launching systems, see MIDAS operational system description, 2-5–2-9.

34. "Summary: June, July, August 1960, Discoverer Project (Research and Development Satellites)," AWF, Admin. Series, box 15, file Gates, Thomas S., Jr., 1959–61(2); letter to Eisenhower from Secretary Neil McElroy, April 30, 1959, WHO, OSS, SUB, DoD Subseries, box 6, file Missiles and Satellites, Vol. 2 (4) [July 1958–January 1959]; Day, "Development and Improvement," 52–59, appendix B, 236.

35. Day et al., *Eye in the Sky*, 72–74; letter to Eisenhower from James H. Douglass, April 11, 1960, AWF, Admin. Series, box 15, file Gates, Thomas S., Jr., 1959–61 (4); NSC 6013: "Status of the United States Military Programs as of 30 June 1960," September 6, 1960, WHO, OSANSA, NSC Series, Status of Projects Subseries, box 9, file NSC 6013, Status of U.S. National Security Programs on June 30, 1960 (1).

36. Letter to the commander ARDC from Brig. Gen. O. J. Ritland, "Subject: WS-117L Program," February 11, 1959, SSB/GCSWS.

37. Letters to General White, chief of staff, from Lieutenant General Schriever, September 15, 1959, SPI, GWU, file "Problems: SAMOS, MIDAS." See also letter to the commander ARDC from Brig. Gen. O. J. Ritland, "Subject: WS-117L Program," February 11, 1959, SSB/GCSWS.

38. Perry, "A History of Satellite Reconnaissance, Vol. 2-A," 35–38.

39. Perry, "A History of Satellite Reconnaissance, Vol. 2-A," 35–38; memorandum for the secretary of the air force, "Subject: Transfer of SAMOS Development Program to the Department of the Air Force," November 17, 1959, and memorandum for the secretary of defense from Joseph V. Charyk, undersecretary of the air force, "Subject: Transfer of the SAMOS, MIDAS and DISCOVERER Development Programs to the Department of the Air Force," February 18, 1960, RG 340, ROSAF, box 367, file 132-59 "Satellite Program Vol. 2"; NASA History Program Office, "The Beginnings of a Division of Labor: Advent and Syncom," www.history.nasa.gov/sp-4102/ch8.htm#214.

40. Herbert York, quoted in Perry, "A History of Satellite Reconnaissance, Vol. 2-A," 40.

41. York refused to approve funding for those aspects of the SAMOS program that he found highly questionable (the readout and ferret systems). Overall funding for FY 1960 amounted to $159.5 million; funding for FY 1961 was less, but the amount is still classified. See Perry, "A History of Satellite Reconnaissance, Vol. 2-A," 20–27, 38–41, 44–45.

42. "Memorandum for the Secretary of Defense, the Special Assistant to the President for Science and Technology, Subject: Presentation on Status of Reconnaissance and Early Warning Satellite Programs," January 28, 1960, WHO, OSANSA, Spec. Assist. Series, Chronological Subseries, box 6, file January–February (1960); "Review of Outer Space Programs under the Auspices of the Department of Defense," May 31, 1960, AWF, NSC Series, box 12, file 446th Meeting of NSC, May 31, 1960; memorandum of conference with the president, February 8, 1960, AWF, DDE Diary Series, box 47, file Staff Notes, February 1960 (2).

43. Memorandum of conference with the president, May 31, 1960 (conference on May 26, 1960), AWF, DDE Diary Series, box 50, file Staff Notes, May 1960 (1).

44. "Memorandum of Conference with the President," May 26, 1960, WHO, OSS, SUB, ALPHA, box 16, file Dr. Kistiakowsky (3); "Letter to Dr. Kistiakowsky from DDE," June 10, 1960, WHO, OSS, SUB, ALPHA, box 16, file Dr. Kistiakowsky (4).

45. "Introductory Remarks by Dr. J. R. Killian Jr.," August 25, 1960, WHO, OSAST, box 15, file Space (July–December 1960) (9); "Memorandum for Attached List," August 19, 1960, WHO, OSANSA, NSC Series, Subject Subseries, box 8, file Reconnaissance Satellites.

46. "The Samos Program," July 29, 1960, WHO, OSAST, box 15, file Space (July–December 1960) (9). For the U.S. Intelligence Board requirements, see also "Samos" (not dated), WHO, OSS, SUB, ALPHA, box 15, file Intelligence Matters (14); memorandum of conference with the president, May 31, 1960, WHO, OSS, SUB, ALPHA, box 16, file Dr. Kistiakowsky (3).

47. The SAMOS briefing notes (probably dated to early July, based on the Intelligence Board requirements that they include) predict that the E-2 camera would fly by April 1961, provide twenty-foot resolution, and satisfy the general search requirements. See "Samos" (not dated).

48. While Edwin Land knew about CORONA, there is no indication that he shared that knowledge with the rest of the panel. "The Samos Program," July 29, 1960, WHO, OSAST, box 15, file Space (July–December 1960) (9).

49. The E-5 and E-6 virtually mimic the DISCOVERER/CORONA configuration, including recovery and flight sequences. See "Samos" (not dated).

50. "The Samos Program," July 29, 1960.

51. Perry, "A History of Satellite Reconnaissance, Vol. 2-A," 55.

52. The official Public Affairs Objectives of September 27, 1960, clearly identified SAMOS as a research and development program only. See "Annex III & IV Public Affairs Objectives," September 27, 1960, WHO, OSANSA, NSC Series, Subject Subseries, box 8, file Reconnaissance Satellites; memorandum for the record, October 1960, WHO, OSS, SUB, ALPHA, box 15, file Intelligence Matters (20).

53. "Reconnaissance Satellite Program Action No. 1-b," September 1, 1960, WHO, NSCSP, Exec. Sect. Subject File Series, box 15, file Reconnaissance Satellites (1960); "Memorandum for the Secretary of Defense, Subject: Reconnaissance Satellite Program," September 1, 1960, WHO, OSS, SUB, ALPHA, box 15, file Intelligence Matters (19); Hall, "The Eisenhower Administration," 67–68; Perry, "A History of Satellite Reconnaissance, Vol. 2-A," 80–86.

54. Hall, "The Eisenhower Administration," 67–68; Richelson, *America's Secret Eyes in Space*, 46–47. The United States now officially acknowledges the National Reconnaissance Office. Providing documents and imagery on satellite operations, their website is www.nro.gov/. Perry, "A History of Satellite Reconnaissance, Vol. 2-A," 61, 84–87.

55. Burrows, *This New Ocean*, 226.

Appendix

1. Warshaw, *Reexamining the Eisenhower Presidency*.

2. To see this point of view it is necessary only to look at some of the most potent authors of the orthodox school: Marquis Childs, *Eisenhower: Captive Hero. A Critical Study of the General and the President* (New York: Harcourt Brace, 1958); Richard Rovere, *Affairs of State: The Eisenhower Years* (New York: Farrar, 1956); Richard Neustadt, *Presidential Power: The Politics of Leadership* (New York: John Wiley & Sons, 1960).

3. Alexander, *Holding the Line*.

4. As an earlier example, see Larson, *Eisenhower*. More recent examples of "revisionist" writings include Greenstein, "Eisenhower"; Rabe, "Eisenhower Revisionism."

5. Rabe, "Eisenhower Revisionism," 97–98.

6. Larson, *Eisenhower*, ix–xii, 12–17; McAuliffe, "Eisenhower, the President," 625–26; Kinnard, *President Eisenhower*, x.

7. Greenstein, "Eisenhower," 575–77, 577–83.

8. Greenstein, "Eisenhower," 579–87, 596–98.

9. McAuliffe, "Eisenhower, the President," 630–31; Ambrose, *Eisenhower: The President*, 9–10, 18–20, 32–35, 624–27.

10. Allen, *Eisenhower*, 1–9.

11. See Immerman, "Confessions," 321–23.

12. Marchetti and Marks, *The CIA*, 7, 76–77, 90, 94.

13. Ranelagh, *The Agency*, 324–25.

14. Andrew, *For the President's Eyes Only*, 3–4, 199–249; Jeffreys-Jones and Andrew, *Eternal Vigilance?*, 83, 85, 90, 92–93, 99.

15. Richelson, *U.S. Intelligence Community*, 36–37.

16. *FEEDBACK Report*, 149–50.

17. Klass, *Secret Sentries*, 1–51, 72–100; Kenden, "U.S. Reconnaissance Satellite Programmes," 243–49; Gaddis, "Transparency"; Burrows, *Deep Black*, 61–2, 82–84, 86–92. Peebles, *Guardians*, 43–70; Richelson, *America's Secret Eyes in Space*, 1–31; Divine, *Sputnik Challenge*, 10–12.

18. McDougall, . . . *the Heavens and the Earth*, 79–110.

19. Executive Order 12951, "Release of Imagery Acquired by Space-Based National Intelligence Reconnaissance Systems," *Federal Register* 60, no. 39 (1995): 10789, in Ruffner, CORONA, 359–90. ARGON and LANYARD were the names for specific camera configurations for CORONA satellites. See Ruffner, "CORONA."

20. Logsdon et al., *Exploring the Unknown*, 1:213–413; Logsdon et al., *Exploring the Unknown*, 2:233–98.

21. Peebles, *Corona Project*, 1–6, 9–14; Day et al., *Eye in the Sky*, 2–3, 29–33, 98–108.

22. Peebles, *Corona Project*, 21–30, 39–52; Day et al., *Eye in the Sky*, 108–13, 41–42.

BIBLIOGRAPHY

Archival Sources

Columbia Center for Oral History, Columbia University, New York, New York
"The Reminiscences of Dr. Richard M. Bissell." June 5, 1967. Interviewed by Ed Edwin.
"The Reminiscences of Dwight D. Eisenhower." July 20, 1967. Interviewed by Ed Edwin.
"The Reminiscences of Andrew J. Goodpaster." April 25, 1967. Interviewed by Ed Edwin.
"The Reminiscences of James R. Killian." November 9, 1969. Interviewed by Stephen White.
"The Reminiscences of Neil McElroy." May 8, 1967. Interviewed by Ed Edwin.
"The Reminiscences of Nathan F. Twining." June 26, 1967. Interviewed by John T. Mason Jr.
Dwight D. Eisenhower Presidential Library, Abilene, Kansas
Dwight D. Eisenhower, Diaries, 1935–38, 1942, 1948–53, 1966, 1968, 1969
Dwight D. Eisenhower, Papers as President of the United States, 1953–61 (Ann Whitman File)
James C. Hagerty, Papers, 1953–61
Bryce N. Harlow, Records, 1953–61
Edward E. "Swede" Hazlett, Papers, 1941–65
"The Reminiscences of Dr. Richard M. Bissell Jr." November 9, 1976. Interviewed by Dr. Thomas Soapes. Dwight D. Eisenhower Presidential Library (OH-382).
"The Reminiscences of General Dwight D. Eisenhower." July 28, 1964. Interviewed by Philip Crowly. Dulles Oral History Project. Dwight D. Eisenhower Presidential Library (OH-14).
"The Reminiscences of John S. D. Eisenhower." March 10, 1972. Interviewed by Dr. Maclyn P. Burg. Dwight D. Eisenhower Presidential Library (OH-15).
"The Reminiscences of Andrew J. Goodpaster." June 26, 1975. Interviewed by Dr. Maclyn P. Burg. Dwight D. Eisenhower Presidential Library (OH-378).

"The Reminiscences of Andrew J. Goodpaster." April 10, 1982. Interviewed
by Malcolm S. McDonald. National Academy for Public Administra-
tion. Eisenhower Presidential Library (OH 477).
"The Reminiscences of Andrew Goodpaster, Ann Whitman, Raymond Saul-
nier, Elmer Staats, Arthur Burns, and Gordon Gray." June 11, 1980.
Interviewed by Hugh Heclo and Anna Nelson. Dwight D. Eisenhower
Presidential Library (OH-508).
"The Reminiscences of George Kistiakowsky." September 17, 1976. Interviewed
by Dr. Thomas Soapes. Dwight D. Eisenhower Presidential Library.
Donald A. Quarles, Papers, 1952–59
U.S. President's Science Advisory Committee, Records, 1957–61
White House Central Files, Official File
White House Office, National Security Council Staff, Papers, 1948–61
White House Office, Office of the Special Assistant for National Security
Affairs, Records, 1952–61
White House Office, Office of the Special Assistant for Science and Technol-
ogy (James R. Killian and George B. Kistiakowsky), Records, 1957–61
White House Office, Office of the Staff Secretary, Records of Paul T. Carroll,
Andrew J. Goodpaster, L. Arthur Minnich, and Christopher H. Russell
Gregg Centre for the Study of War and Society, University of New Brunswick,
Canada
Harriet Irving Library, University of New Brunswick, Canada
Library of Congress, Washington DC
Thomas D. White, Papers
National Archives and Records Administration II, College Park, Maryland
RG 59: General Records of the Department of State
RG 340: Records of the Office of the Secretary of the Air Force
Perry, Robert L. "A History of Satellite Reconnaissance, Vol. 1: CORONA." National
Reconnaissance Office, 1964; revised 1967, 1972, 1973. Dwayne Day Collection.
———. "A History of Satellite Reconnaissance, Vol. 2-A: SAMOS." National Recon-
naissance Office, 1963–65; revised 1972, 1973.
———. "A History of Satellite Reconnaissance, Vol. 4: NRO History." National
Reconnaissance Office (undated rough draft).
———. "A History of Satellite Reconnaissance, Vol. 5: Management of the National
Reconnaissance Program, 1960–1965." National Reconnaissance Office, 1969.
RAND Corporation, Santa Monica, California
Augustein, Bruno W. "Evolution of U.S. Military Space Program, 1945–
1960: Some Key Events in Study, Planning, and Program Develop-
ment." P-6814. September 1982.
———. "Policy Analysis in the National Space Program." P-4137. July 1969.
———. "A Revised Development Program for a Ballistic Missile of Inter-
continental Range." SM-21. February 8, 1954.

Davies, M. E., and A. H. Katz. "Comments on T. I. Edwards's 'New Requirements for Photo Reconnaissance.'" D-3429. January 20, 1956.

Davies, Merton E., and William R. Harris. *RAND's Role in the Evolution of Balloon and Satellite Observation Systems and Related US Space Technology.* RAND Publication Series R-3692-RC. September 1988.

Katz, Amrom H. "Recommendations, Formal and Informal, on Reconnaissance Satellite Matters." D(L)-6421 SOFS-RECCE-5. May 11, 1959.

———. "Soviet Requirements for a Reconnaissance Satellite." D-4883. January 23, 1958.

Katz, Amrom H., and M. E. Davies. "On the Utility of Very Large Satellite Payloads for Reconnaissance." D-5817. November 14, 1958.

Ware, Willis H. *Project RAND and Air Force Decision Making.* P-5737. November 1976.

Space Policy Institute, George Washington University, Washington DC

Published Sources

Adams, Sherman. *Firsthand Report: The Story of the Eisenhower Administration.* New York: Harper & Brothers, 1961.

"Administration Divided on Sputnik." *Aviation Week* 67 (October 1957): 30.

Aid, Matthew M. "The National Security Agency and the Cold War." *Intelligence and National Security* 16 (Spring 2001): 27–66.

Aid, Matthew M., and Cees Wiebes. "Introduction: The Importance of Signals Intelligence in the Cold War." *Intelligence and National Security* 16 (Spring 2001): 1–26.

Albats, Yeugenia. *KGB: State within a State.* London: I. B. Tauris, 1994.

Albertson, Dean. *Eisenhower as President.* New York: Hill and Wang, 1963.

Alexander, Charles C. *Holding the Line: The Eisenhower Era 1952–1961.* Bloomington: Indiana University Press, 1975.

Aliano, Richard A. *American Defense Policy from Eisenhower to Kennedy.* Athens: Ohio University Press, 1975.

Allen, Craig. *Eisenhower and the Mass Media: Peace, Prosperity, and Prime-Time TV.* Chapel Hill: University of North Carolina Press, 1993.

Ambrose, Stephen E. *Eisenhower.* Vol. 1: *Soldier, General of the Army, President-Elect, 1890–1952.* New York: Simon & Schuster, 1983.

———. *Eisenhower.* Vol. 2: *The President.* New York: Simon and Schuster, 1983–84.

———. *Eisenhower: Soldier and President.* New York: Simon & Schuster, 1990.

———. *Ike's Spies: Eisenhower and the Espionage Establishment.* New York: Doubleday, 1981.

———. *Rise to Globalism: American Foreign Policy since 1938.* Baltimore: Penguin, 1971.

Ambrose, Stephen E., and James Alden Barber. *The Military and American Society: Essays and Readings.* New York: Free Press, 1972.

"American Astronomers Report." *Sky and Telescope* 15 (January 1956): 112–14.

Anderson, Frank W., Jr. *Orders of Magnitude: A History of NACA and NASA, 1915–1980.* 2nd ed. Washington DC: Government Printing Office, 1981.

Anderton, David A. "Details of Sputnik Surprise Scientists." *Aviation Week* 67 (October): 30–31.

——. "Midas Tests Detection System." *Aviation Week* 72 (May 1960): 36.

"And in Four Years?" *New Republic* 137 (October 1957): 3–6.

Andrew, Christopher. *For the President's Eyes Only: Secret Intelligence and the American Presidency from Washington to Bush.* New York: Harper Collins, 1995.

Andrew, Christopher, and David Dilks, eds. *The Missing Dimension: Governments and Intelligence Communities in the Twentieth Century.* London: Macmillan, 1985.

"Another Russian Satellite." *Sky and Telescope* 17 (December 1959): 55, 65.

Armacost, Michael. *The Politics of Weapons Innovation: The Thor-Jupiter Controversy.* New York: Columbia University Press, 1969.

Aronsen, Lawrence. "Seeing Red: U.S. Air Force Assessments of the Soviet Union, 1945–1949." *Intelligence and National Security* 16 (Summer 2001): 103–32.

Bainbridge, William Sims. *The Space Flight Revolution: A Sociological Study.* New York: John Wiley and Sons, 1976.

Ball, Desmond, and Jeffrey Richelson, eds. *Strategic Nuclear Targeting.* Ithaca NY: Cornell University Press, 1986.

Bamford, James. *Body of Secrets: Anatomy of the Ultra-Secret National Security Agency.* New York: Knopf, 2002.

Barnet, Richard J. *The Economy of Death.* New York: Atheneum, 1970.

Barnford, James. *The Puzzle Palace: A Report on America's Most Secret Agency.* New York: Penguin, 1985.

Barrow, John. *KGB: The Secret Work of Soviet Agents.* New York: Reader's Digest Press, 1974.

Baucom, Donald R. *The Origins of SDI, 1944–1983.* Lawrence: University Press of Kansas, 1992.

Beard, Edmund. *Developing the ICBM: A Study in Bureaucratic Politics.* New York: Columbia University Press, 1976.

Bechhoefer, Bernhard G. *Postwar Negotiations for Arms Control.* Washington DC: Brookings Institution, 1961.

Becker, William H. *The Dynamics of Business-Government Relations: Industry and Exports 1893–1921.* Chicago: University of Chicago Press, 1982.

Bell, Coral. *Negotiations from Strength: A Study in the Politics of Power.* New York: Knopf, 1963.

Benko, Marietta, et al. *Space Law in the United Nations.* Boston: Martinus Nijhoff, 1985.

Benson, Robert Louis, and Michael Warner, eds. *VENONA: Soviet Espionage and the American Response, 1939–1957.* Washington DC: National Security Agency and Central Intelligence Agency, 1997.

Berding, Andrew. *Dulles on Diplomacy*. Princeton NJ: VanNostrand, 1965.

Bergaust, Erik, and William Beller. *Satellite!* Garden City NY: Hanover Huge, 1956.

Berger, Carl. *The Air Force in Space, Fiscal Year 1961*. Washington DC: USAF Historical Division Liaison Office, 1966.

Bernstein, Barton J. "The Challenges and Dangers of Nuclear Weapons: Foreign Policy and Strategy, 1941–1978." *Maryland Historian* 9 (Spring 1978): 73–99.

——. "Foreign Policy in the Eisenhower Administration." *Foreign Service Journal* 50 (May 1973): 17–38.

Beschloss, Michael R. *May-Day: Eisenhower, Khrushchev and the U-2 Affair*. New York: Harper & Row, 1986.

Best, G., and A. Wheatcroft, eds. *War, Economy and the Military Mind*. London: Croom Helm, 1976.

Biddle, Tami Davis. "Handling the Soviet Threat: 'Project Control' and the Debate on American Strategy in the Early Cold War Years." *Journal of Strategic Studies* 12 (September 1989): 273–302.

Bilstein, Roger E. *Orders of Magnitude: A History of the NACA and NASA, 1915–1990*. Washington DC: NASA, Office of Management, Scientific and Technical Information Division, 1989.

Bissell, Richard M., Jr. "Origins of the U-2." *Air Power History* 36 (Winter 1989): 15–24.

Bissell, Richard M., Jr., Jonathan E. Lewis, and Francis T. Pudlo. *Reflections of a Cold Warrior: From Yalta to the Bay of Pigs*. New Haven CT: Yale University Press, 1996.

Blakeslee, Alton L. "Sphere 'Space Columbus.'" *Windsor (Ontario) Daily Star*, October 5, 1957.

Bottome, Edgar M. *The Missile Gap: A Study in the Formulation of Military Policy*. Plainsboro NJ: Associated University Press, 1971.

Bowen, Lee. *The Threshold of Space: The Air Force in the National Space Program, 1945–1959*. Washington DC: USAF Historical Division Liaison Office, 1960.

Bowie, Robert R., and Richard H. Immerman. *Waging Peace: The National Security Policies of the Eisenhower Administration*. New York: Oxford University Press, 1998.

Boxhall, Peter. "Aerial Photography and Photographic Interpreters: 1915 to the Gulf War." *Army Quarterly and Defence Journal* (Great Britain) 122, no. 2 (1992): 204–9.

Branyan, Robert L., and Lawrence H. Larsen, eds. *The Eisenhower Administration, 1953–1961: A Documentary History*. 2 vols. New York: Random House, 1971.

Breuer, William B. *Race to the Moon: America's Duel with the Soviets*. Westport CT: Praeger, 1993.

Broadwater, Jeff. *Eisenhower and the Anti-Communist Crusade*. Chapel Hill: University of North Carolina Press, 1992.

Brodie, Bernard, ed. *The Absolute Weapon: Atomic Power and World Order*. New York: Harcourt, Brace, 1946.

———. *The Atomic Bomb and American Security*. New Haven CT: Yale University Press, 1945.

———. *Strategy and National Interests: Reflections for the Future*. New York: National Strategy Information Center, 1971.

———. *Strategy in the Missile Age*. Princeton NJ: Princeton University Press, 1967.

———. *War and Politics*. New York: Macmillan, 1973.

Brodie, Bernard, and Fawn Brodie. *From Crossbow to H-Bomb*. Bloomington: Indiana University Press, 1973.

Brodie, Bernard, Michael D. Intriligator, and Roman Kolkowicz, eds. *National Security and International Stability*. Cambridge MA: Oelgeschlager, Gunn & Hain, 1983.

Brooks, Courtney G., James M. Grimwood, and Loyd S. Swensen Jr. *Charter for Apollo: A History of Manned Lunar Spacecraft*. Washington DC: NASA Historical Series, NASA SP-4205, 1975.

Brown, S. *The Faces of Power: Constancy and Change in Untied States Foreign Policy from Truman to Johnson*. New York: Columbia University Press, 1968.

Brownell, George A. *The Origin and Development of the National Security Agency*. Walnut Creek CA: Aegean Park Press, 1981.

Brugioni, Dino A. *Eyeball to Eyeball*. New York: Random House, 1991.

Brune, Lester F. *The Origins of American National Security Policy: Sea Power, Air Power and Foreign Policy*. Manhattan KS: MA/AH, 1981.

Bulkeley, Rip. *The Sputnik Crisis and Early U.S. Space Policy: A Critique of the Historiography of Space*. Bloomington: Indiana University Press, 1991.

Burgess, Eric. *Long Range Ballistic Missiles*. New York: Macmillan, 1962.

Burk, Kathleen, and Melvyn Stokes, eds. *The United States and the European Alliance since 1945*. New York: Berg, 1999.

Burk, Robert Fredrick. *Dwight D. Eisenhower: Hero and Politician*. Boston: Twayne, 1986.

Burrows, William E. *Deep Black: Space Espionage and National Security*. New York: Random House, 1986.

———. *This New Ocean: The Story of the First Space Age*. New York: Random House, 1998.

Bush, Vannevar. *Modern Arms and Free Men: A Discussion of the Role of Science in Preserving Democracy*. New York: Simon & Schuster, 1949.

Capitanchik, David. *The Eisenhower Presidency and American Foreign Policy*. New York: Humanities Press, 1969.

Capitanchik, David, and Richard D. Eichenberg. *Defence and Public Opinion*. Boston: Routledge & K. Paul, 1983.

Carey, Olmer L. *The Military-Industrial Complex and U.S. Foreign Policy*. Pullman: Washington State University Press, 1969.

Caridi, Ronald J. *20th Century American Foreign Policy: Security and Self-Interest*. Englewood Cliffs NJ: Prentice-Hall, 1974.

Carlton, David, and Carlo Schaerf, eds. *Arms Control and Technological Innovation*. London: Croom Helm, 1977.

Carter, Dale. *The Rise and Fall of the American Rocket State*. London: Verso Press, 1988.

Carter, Paul A. *Politics, Religion and Rockets: Essays in 20th Century American History*. Tucson: University of Arizona Press, 1991.

Charters, David A. "Nine Days in May: The U-2 Deception." In *Deception Operations: Studies in the East-West Context*, edited by David A. Charters and Maurice A. J. Tugwell, 333–50. London: Brassey's, 1990.

Christol, Carl Q. *Space Law: Past Present and Future*. Cambridge: Eluwer Law & Taxation, 1990.

Clark, Evert. "ARPA Reveals Rocket Satellite Programs." *Aviation Week* 69 (November 1958): 26–27.

Clark, John D. *Ignition! An Informal History of Liquid Rocket Propellants*. New Brunswick NJ: Rutgers University Press, 1972.

Clarke, A. C., ed. *The Coming of the Space Age*. New York: Meredith Press, 1967.

Clausen, Henry C., and Bruce Lee. *Pearl Harbor: Final Judgement*. New York: Crown, 1992.

Clayton, James L. *The Economic Impact of the Cold War: Sources and Readings*. New York: Harcourt, Brace & World, 1970.

Cline, Ray S. *Secrets, Spies and Scholars: Blue Print of the Essential CIA*. Washington DC: Acropolis Books, 1976.

Collins, Martin J., and Sylvia K. Fries, eds. *A Space Faring Nation: Perspectives on American Space History and Policy*. Washington DC: Smithsonian Institute Press, 1991.

Collins, Martin J., and Sylvia K. Kraemer, eds. *Space: Discovery and Exploration*. Southport CT: Simon and Schuster for Hugh Lauter Levin Associates, 1993.

Combs, Jerald A. *The History of American Foreign Policy*. New York: McGraw-Hill, 1997.

Compton, James V., ed. *America and the Origins of the Cold War*. Boston: Houghton Mifflin, 1972.

Condit, Doris M. *History of the Office of the Secretary of Defense*. Vol. 2 of *The Test of War, 1950–1953*. Washington DC: Historical Office of the Secretary of Defense, 1988.

Condit, Kenneth W. *History of the Joint Chiefs of Staff*. Vol. 6: *The Joint Chiefs of Staff and National Policy, 1955–1956*. Washington DC: Historical Office, Government Printing Office, 1992.

Congress and the Nation: A Review of Government and Politics in the Post War Years. Washington DC: Congressional Quarterly Services, 1965.

Cook, Blanche Wiesen. *The Declassified Eisenhower: A Divided Legacy*. Garden City NY: Doubleday, 1981.

Cook, Fred J. *The U-2 Incident, May 1960: An American Spy Plane Downed over Russia Intensifies the Cold War*. New York: Watts, 1973.

———. *The Warfare State*. New York: Macmillan, 1962.

Coolbaugh, James S. "Genesis of the USAF's First Satellite Programme." *Journal of the British Interplanetary Society* 51 (1998): 283–300.

Cooling, Benjamin Franklin, ed. *War, Business, and American Society: Historical Perspectives in the Military-Industrial Complex*. Port Washington NY: Kennikat Press, 1977.

Cornett, Lloyd H., and Frederick I. Ordway III, eds. *History of Rocketry and Astronautics*. AAS History Series 9. San Diego CA: American Astronautical Society, 1989.

Corson, William R., and Robert T. Crowley. *The New KGB: Engine of Secret Power*. New York: William Morrow, 1986.

Cox, Donald W. *The Space Race: From Sputnik to Apollo . . . and Beyond*. New York: Chilton Books, 1962.

Cunningham, William G. *The Aircraft Industry: A Study in Industrial Location*. Berkeley: University of California Press, 1979.

Damus, Richard V. "Scientists and Statesmen: Eisenhower's Science Advisers and National Security Policy 1953–1961." PhD dissertation, Ohio State University, 1993. University Microfilms No. 93-25485.

Daniloff, Nicholas. *The Kremlin and the Cosmos*. New York: Knopf, 1972.

Darling, Arthur B. *The Central Intelligence Agency: An Instrument of Government to 1950*. University Park: Pennsylvania State University Press, 1990.

"Data from Satellites Sent by Two Methods." *Science News Letter* 70 (February 1956): 280.

Davis, Kenneth Sydney. *Eisenhower, American Hero: The Historical Record of His Life*. New York: American Heritage, 1969.

Dawson, Virginia P. *Engines and Innovation: Lewis Laboratory and American Propulsion Technology*. Washington DC: NASA Office of Management, Scientific and Technical Information Division, 1991.

Day, Dwayne A. "Listening from Above: The First Signals Intelligence Satellite." *Spaceflight* 41 (August 1999): 339–46.

———. "Paradigm Lost." *Space Policy* 11, no. 3 (1995): 153–59.

Day, Dwayne A., John M. Logsdon, and Brian Latell. *Eye in the Sky: The Story of the Corona Spy Satellites*. Washington DC: Smithsonian Institution Press, 1998.

DeGarmont, Sanche. *The Secret War: The Story of International Espionage since World War II*. London: Andre Deutsch, 1962.

"Democratic Senators Demand a Full Investigation of the Nation's Satellite Program." *New York Times*, October 8, 1957.

de Silva, Peer. *Sub Rosa: The CIA and the Use of Intelligence*. New York: Times Books, 1978.

Dethloff, Henry C. *Suddenly, Tomorrow Came: A History of the Johnson Space Center*. NASA SP-4307. Washington DC: NASA, 1993.

Devorkin, David H. *Science with a Vengeance: How the Military Created the US Space Sciences after World War II*. New York: Springer-Verlag, 1992.

"Discoverer Failure Caused by Inverter." *Aviation Week* 71 (November 1959): 33.

"Discoverer Satellite Enters Orbit; Payload Reaches 537 Mi. Apogee." *Aviation Week* 71 (August 1959): 36.

Divine, Robert A. *American Foreign Policy since 1945.* Chicago: Quadrangle Books, 1969.

———. *Eisenhower and the Cold War.* Oxford: Oxford University Press, 1981.

———. *Foreign Policy and U.S. Presidential Elections 1952–1960.* New York: New View Points, 1974.

———. *Since 1945: Politics and Diplomacy in Recent American History.* New York: John Wiley & Sons, 1979.

———. *The Sputnik Challenge: Eisenhower's Response to the Soviet Satellite.* New York: Oxford University Press, 1993.

Domhoff, G. William, and Hoyt B. Ballard, eds. *C. Wright Mills and the Power Elite.* Boston: Beacon Press, 1969.

Donovan, John C. *The Cold Warriors: A Policy Making Elite.* Lexington M A: Heath, 1974.

———. *The Policy Makers.* New York: Western, 1970.

Donovan, Robert J. *Conflict and Crisis: The Presidency of Harry S. Truman, 1945–1948.* New York: Norton, 1977.

———. *Eisenhower: The Inside Story.* New York: Harper & Brothers, 1956.

Dornberger, W. R., and S. F. Slinger. "Can U.S. Still Win Missile Race?" *U.S. News & World Report,* November 15, 1957, 52–108.

Dulles, Allan. *The Craft of Intelligence.* New York: Harper & Row, 1963.

Durch, William J. *National Interests and the Military Use of Space.* Cambridge M A: Ballinger, 1984.

Economic Report of the President. January 20, 1955. Washington D C: Government Printing Office, 1955.

Eisenhower, Dwight D. *At Ease: Stories I Tell to Friends.* Garden City N Y: Doubleday, 1967.

———. *The White House Years.* Vol. 1: *Mandate for Change.* Garden City N Y: Doubleday, 1963.

———. *The White House Years.* Vol. 2: *Waging Peace.* Garden City N Y: Doubleday, 1965.

Eisenhower, Dwight D., and Robert H. Ferrell. *Eisenhower Diaries.* New York: Norton, 1981.

Eisenhower, Dwight D., Everett E. Hazlett, and Robert Griffith. *Ike's Letters to a Friend, 1941–1958.* Lawrence: University Press of Kansas, 1984.

Eisenhower, John. *Strictly Personal.* Garden City N Y: Doubleday, 1974.

"Eisenhower Holds Conferences on Russian Satellite; Navy Officer Says Soviets Have Medium Range Missile." *Wall Street Journal,* October 9, 1957.

Elder, Donald Cameron, III. "Out from Behind the Eight-Ball: Echo I and the Emergence of the American Space Program, 1957–60." PhD dissertation, University of California, San Diego, 1989. D A I 1990 50(8): 2620-A. D A 8926240.

Elliott, Derek Wesley. "Finding an Appropriate Commitment: Space Policy Development under Eisenhower and Kennedy, 1954–63." PhD dissertation, George Washington University, 1992. DAI 1992 53(4): 1252-A. DA9224235.

Enke, Stephen. *Defense Management*. Englewood Cliffs NJ: Prentice-Hall, 1967.

Ewald, William. *Eisenhower the President: Crucial Days, 1951–1960*. Englewood Cliffs NJ: Prentice-Hall, 1981.

"Eyes-in-the-Skies: Peace and War." *Newsweek*, December 1958, 63–64.

Falk, Stanley L. "The National Security Council under Truman, Eisenhower, and Kennedy." *Political Science Quarterly* 79 (September 1964): 403–34.

Fawcett, J. E. S. *International Law and the Uses of Outer Space*. New York: Oceana, 1968.

Federal Economic Policy, 1945–1965. Washington DC: Congressional Quarterly, 1966.

Ferrell, Robert H. *America in a Divided World, 1945–1972*. Columbia: University of South Carolina Press, 1975.

Ferris, John. "Coming in from the Cold War: The Historiography of American Intelligence, 1945–1990." *Diplomatic History* 19 (Winter 1995): 87–115.

Fox, William T. R. "Representativeness and Efficiency: Dual Problem of Civil-Military Relations." *Political Science Quarterly* 76 (June 1961): 354–66.

Freedman, Lawrence. *The Evolution of Nuclear Strategy*. London: Macmillan, 1983.

———. *U.S. Intelligence and the Soviet Strategic Threat*. Boulder CO: Westview Press, 1977.

Freeman, Eva C. *MIT Lincoln Laboratory: Technology in the National Interest*. Lexington MA: MIT Press, 1995.

Friedberg, Aaron L. "Review Essay: Science, the Cold War, and the American State." *Diplomatic History* 20 (Winter 1996): 107–18.

Frutkin, Arnold W. *International Cooperation in Space*. Englewood Cliffs NJ: Prentice-Hall, 1965.

Futrell, Robert Frank. *Ideas, Concepts, Doctrine*. Vol. 1: *A History of Basic Thinking in the United States Air Force, 1907–1960*. Maxwell AFB AL: Air University Press, 1989.

———. *Ideas, Concepts, Doctrine*. Vol. 2: *A History of Basic Thinking in the United States Air Force, 1961–1984*. Maxwell AFB AL: Air University Press, 1989.

Gaddis, John Lewis. *The Long Peace*. New York: Oxford University Press, 1987.

———. *Strategies of Containment: A Critical Appraisal of Post War U.S. National Security Policy*. New York: Oxford University Press, 1982.

———. *The United States and the Origins of the Cold War: 1941–1947*. New York: Columbia University Press, 1972.

Gal, Gyula. *Space Law*. New York: Oceana, 1969.

Galloway, E. *Legal Problems of Space Exploration: A Symposium*. Prepared for the Use of the Committee on Aeronautics and Space Sciences, U.S. Senate. Washington DC: Government Printing Office, 1961.

Gamson, William, and Andre Modigliani. *Untangling the Cold War: A Strategy for Testing Rival Theories*. Boston: Little, Brown, 1971.

Gansler, Jacques S. *The Defense Industry.* Cambridge MA: MIT Press, 1980.

Gavin, James M. *War and Peace in the Space Age.* New York: Harper and Brothers, 1958.

George, A. L., and R. Smoke. *Deterrence in American Foreign Policy: Theory and Practice.* New York: Columbia University Press, 1974.

George, Alexander L. *Presidential Decision Making in Foreign Policy: The Effective Use of Information and Advice.* Boulder CO: Westview Press, 1980.

Giancoli, Douglas C. *Physics.* Englewood Cliffs NJ: Prentice Hall, 1985.

Gibney, Frank B., and George J. Feldman. *The Reluctant Space Farers: A Study in the Politics of Discovery.* New York: New American Library of World Literature, 1968.

Gilpin, Robert. *American Scientists and Nuclear Weapons Policy.* Princeton NJ: Princeton University Press, 1962.

———. *War and Change in World Politics.* Cambridge: Cambridge University Press, 1981.

Gilpin, Robert, and Christopher Wright, eds. *Scientists and National Policy Making.* New York: Columbia University Press, 1964.

Goddard, George W., and DeWitt S. Cop. *Overview: A Lifelong Adventure in Aerial Photography.* Garden City NY: Doubleday, 1969.

Goldberg, Alfred. *History of the Office of the Secretary of Defense.* Washington DC: Historical Office, Office of the Secretary of Defense, U.S. Government Printing Office, 1988.

———. *A History of the U.S. Air Force, 1907–1957.* Princeton NJ: Van Nostrand, 1957.

Goldman, Eric F. *The Crucial Decade—and After: America, 1945–60.* New York: Random House, 1960.

Goldman, Nathan C. *American Space Law: International and Domestic.* Ames: Iowa State University Press, 1988.

———. *Space Policy: An Introduction.* Ames: Iowa State University Press, 1992.

Goldstein, Martin E. *Arms Control and Military Preparedness from Truman to Bush.* New York: Peter Lang, 1993.

Goodpaster, Andrew J. *For the Common Defense.* Lexington MA: D. C. Heath, 1977.

Gorgol, John Francis. *The Military-Industrial Firm: A Practical Theory and Model.* New York: Praeger, 1972.

———. "A Theory of the Military-Industrial Firm." PhD dissertation, Columbia University, 1969.

Gorn, Michael H., ed. *Prophesy Fulfilled: "Towards New Horizons" and Its Legacy.* Washington DC: Air Force History and Museum Programs, 1994.

Gosnell, Harold F. *Truman's Crisis: A Political Biography of Harry S. Truman.* Westport CT: Greenwood Press, 1980.

Graham, Daniel Orrin. *U.S. Intelligence at the Crossroads.* Intelligence Requirements for the 1980s, No. 1. Washington DC: National Strategy Information Center, 1979.

Graham, Ron, and Roger E. Read. *Manual of Aerial Photography.* Boston: Butterworth, 1986.

Gray, Colin S. *The Soviet-American Arms Race.* Westmead, Farmborough, England: D. C. Heath, 1976.

Gray, George C. *Frontiers of Flight: The Story of NACA Research.* New York: Knopf, 1948.

Green, Constance M., and Milton Lomask. *Vanguard: A History.* NASA SP-4202. Washington DC: Smithsonian Institute Press, 1971.

Greenberg, Daniel S. *The Politics of Pure Science.* New York: New American Library, 1968.

Greenstein, Fred I. "Eisenhower as an Activist President: A Look at New Evidence." *Political Science Quarterly* 94 (Winter 1979–80): 575–99.

Greenwood, Ted. *Reconnaissance, Surveillance, and Arms Control.* London: International Institute for Strategic Studies, 1972.

Greenwood, Ted, Harold A. Feiveson, and Theodore B. Taylor. *Nuclear Proliferation: Motivations, Capabilities and Strategies for Control.* New York: McGraw-Hill, 1976.

Griffith, Robert, ed. *Major Problems in American History since 1945.* Lexington MA: D. C. Heath, 1992.

Grose, Peter. *Gentleman Spy: The Life of Allen Dulles.* New York: Houghton Mifflin, 1994.

Gruntman, Mike. *Blazing the Trail: The Early History of Spacecraft and Rocketry.* Reston VA: Institute of Aeronautics and Astronautics, 2004.

Hagerty, James C. "Plans for Launching of Earth-Circling Satellites." *Department of State Bulletin* 63 (August 1955); 218.

Halberstam, David. *The Fifties.* New York: Ballantine Books, 1993.

Hall, R. Cargill. "The Eisenhower Administration and the Cold War: Framing American Astronautics to Serve National Security." *Prologue* 27 (Spring 1995): 59–73.

———, ed. *Essays on the History of Rocketry and Astronautics: Proceedings of the Third through Sixth History Symposia of the International Academy of Astronautics.* 2 vols. Washington DC: NASA Conference Publication 2014, 1977.

———. "From Concept to National Policy: Strategic Reconnaissance in the Cold War." *Prologue* 28 (Summer 1996): 106–25.

Hallion, Richard P. *On the Frontier: Flight Research at Dryden, 1946–1981.* Washington DC: NASA, 1984.

Hansen, James R. *Engineer in Charge: A History of the Langley Aeronautical Lab, 1917–1958.* Washington DC: NASA, 1987.

Harquhar, John T. "A Need to Know: The Role of Air Force Reconnaissance in War Planning, 1945–1953." PhD dissertation, Ohio State University, 1991.

Harris, Seymour E. *The Economics of the Political Parties.* New York: Macmillan, 1962.

Hartt, Julian. *Mighty Thor.* New York: Duell, Sloan and Pearce, 1961.

Hawkes, Russell. "USAF's Satellite Test Center Grows." *Aviation Week and Space Technology*, May 30, 1960, 57–63.

Hechler, Ken. *The Endless Space Frontier: A History of the House Committee on Science and Astronautics, 1959–1978.* AAS History Series, vol. 4. San Diego CA: American Astronautical Society, 1982.

Heiman, Grover. *Aerial Photography: The Story of Aerial Mapping and Reconnaissance.* New York: Macmillan, 1972.

Heller, Francis H., ed. *Economics and the Truman Administration.* Lawrence: Regent Press of Kansas, 1981.

————. *The Truman White House: The Administration of the Presidency, 1945–1953.* Lawrence: Regent Press of Kansas, 1980.

Heppenheimer, T. A. *Countdown: A History of Space Flight.* New York: John Wiley & Sons, 1997.

Herken, Gregg. *Cardinal Choices: Presidential Science Advisory from the Atomic Bomb to SDI.* New York: Oxford University Press, 1992.

Herman, Michael. *Intelligence Power in Peace and War.* Cambridge: Cambridge University Press, 1996.

Hethloff, Henry C. *Suddenly Tomorrow Came . . . A History of the Johnson Space Center.* Washington DC: NASA, 1993.

Hewlett, Richard G., and Oscar E. Anderson Jr. *A History of the United States Atomic Energy Commission.* Vol. 1: *The New World, 1939–1945.* University Park: Pennsylvania State University Press, 1962.

Hirsch, Richard, and Joseph Trento. *The National Aeronautics and Space Administration.* New York: Praeger, 1973.

Hochman, Sandra, and Sybil Wong. *Satellite Spies: The Frightening Impact of a New Technology: An Investigation.* Indianapolis: Bobbs-Merrill, 1976.

Hogan, Michael J. *A Cross of Iron: Harry S. Truman and the Origins of the National Security State, 1945–1954.* New York: Cambridge University Press, 1998.

Holloway, David. *Stalin and the Bomb.* New York: Yale University Press, 1994.

Holt, Pat M. *Secret Intelligence and Public Policy: A Dilemma of Democracy.* Washington DC: Congressional Quarterly, 1995.

Hoopes, Townsend. *The Devil and John Foster Dulles.* Boston: Little, Brown, 1973.

Hopple, Gerald W., and Bruce W. Watson, eds. *The Military Intelligence Community.* Boulder CO: Westview Press, 1986.

Horowitz, David, ed. *Corporations of the Cold War.* New York: Monthly Review Press, 1969.

Horowitz, I. L. *Politics and People.* New York: Ballantine Books, 1962.

Hotz, Robert. "Top Priority for Military Space Systems." *Aviation Week* 72 (May 1960): 2.

Hughes, Emmet John. *America the Invincible: A brief inquiry, written with much anxiety and some anger, dealing with matters of foreign policy, seeking probable sources of our peril and causes of our fear, as well as the suggestion of possible ways by which we might, while enduring the stern trial, hold hope, tempered by reason, to live in freedom more profound and peace less precarious.* Garden City NY: Doubleday, 1959.

———. *The Ordeal of Power: A Political Memoir of the Eisenhower Years.* New York: Atheneum, 1963.

Hunley, J. D., ed. *The Birth of NASA: The Diary of T. Keith Glennan.* Washington DC: National Aeronautics and Space Administration, NASA Historical Office, 1993.

Hunt, Linda. *Secret Agenda: The United States Government, Nazi Scientists and Project Paperclip, 1945–1990.* New York: St. Martin's Press, 1991.

Huntington, Samuel P. *Changing Patterns of Military Politics.* New York: Free Press of Glencoe, 1962.

———. *The Common Defense: Strategic Programs in National Politics.* New York: Columbia University Press, 1961.

"Ike, Cabinet Discuss Missiles." *Windsor (Ontario) Daily Star,* October 11, 1957.

Immerman, Richard H. "Confessions of an Eisenhower Revisionist: An Agonizing Reappraisal." Bernath Lecture. *Diplomatic History* 14 (Summer 1990): 310–42.

———. "The United States and the Geneva Conference of 1954: A New Look." *Diplomatic History* 14 (Winter 1990): 43–66.

Inderfurth, Karl F., and Loch K. Johnson. *Decisions of the Highest Order: Perspectives on the National Security Council.* Belmont CA: Brooks/Cole, 1988.

Jasentuliyama, Nandasiri. *International Space Law and the United Nations.* Boston: Kluwer Law International, 1999.

Javits, Jacob K., Charles J. Hitch, and Arthur F. Burns. *The Defense Sector and the American Economy.* New York: New York University Press, 1968.

Jeffreys-Jones, Rhodri. *The CIA and American Democracy.* New Haven CT: Yale University Press, 2003.

Jeffreys-Jones, Rhodri, and Christopher Andrew. *Eternal Vigilance? 50 Years of the CIA.* Portland OR: Frank Cass, 1997.

Jeszenszky, Elizabeth. "The Frustrating Evolution of the Defense Intelligence Agency." MA thesis, American University, 1992. AAC 1353496.

Johnsen, Katherine. "Senate Group Probes Satellite Progress." *Aviation Week* 67 (October 1957): 31–32.

Johnson, Clarence L. "Kelly," and Maggie Smith. *Kelly: More Than My Share of It All.* Washington DC: Smithsonian Institution Press, 1985.

Johnson, Dana J., Scott Pace, and C. Bryan Goddard. *Space: Emerging Options for National Power.* Santa Monica CA: RAND, 1988.

Johnson, Gerald W. "Eisenhower and Truman." *New Republic* 137 (October 1957): 9.

Johnson, Lock K. *America's Secret Power: The CIA in a Democratic Society.* New York: Oxford University Press, 1989.

———. *Bombs, Bugs, Drugs, and Thugs: Intelligence and America's Quest for Security.* New York: New York University Press, 2000.

Joseph, James. "Explorer III: Our Spy in Outer Space." *Popular Mechanics* 109, no. 4 (1958): 81–86, 248, 250, 266, 268.

Kam, Ephrair. *Surprise Attack: The Victims' Perspective.* Cambridge MA: Harvard University Press, 1988.

Kaplan, Fred M. *Dubious Spector: A Second Look at the "Soviet Threat."* Pamphlet No. 6. Washington DC: Transnational Institute, 1977.

———. *The Wizards of Armageddon.* Stanford: Stanford University Press, 1991.

Karas, Thomas. *The New High Ground: Systems and Weapons of Space Age War.* New York: Simon & Schuster, 1983.

Katz, James Everett. *Presidential Politics and Science Policy.* New York: Praeger, 1978.

Keefer, Edward C. "President Dwight D. Eisenhower and the End of the Korean War." *Diplomatic History* 10 (Summer 1986): 267–89.

Kenden, Anthony. "U.S. Reconnaissance Satellite Programmes." *Spaceflight* 20 (July 1978): 243–62.

Kennan, George F. "The Sources of Soviet Conduct." *Foreign Affairs* 65 (Spring 1987): 852–68.

Kennedy, Gavin. *The Economics of Defense.* London: Faber & Faber, 1975.

Kevles, Daniel J. "Scientists, the Military, and the Control of Postwar Defense Research: The Case of the Research Board for National Security 1944–1946." *Technology and Culture* 16 (January 1975): 20–47.

Khrushchev, N. S. *Khrushchev Remembers: The Last Testament.* Trans. Strobe Talbott. Boston: Little, Brown, 1974.

"Khrushchev Asks World Rule of the Satellite and Missiles If Part of Wide U.S.-Soviet Pact." *New York Times,* October 8, 1957.

Killian, James R. *The Education of a College President: A Memoir.* Cambridge MA: MIT Press, 1985.

———. *Sputniks, Scientists, and Eisenhower: A Memoir of the First Special Assistant to the President for Science and Technology.* Cambridge MA: MIT Press, 1977.

King-Hele, Desmond. *Observing Earth Satellites.* New York: Van Nostrand Reinhold, 1983.

———. *A Tapestry of Orbits.* Cambridge: Cambridge University Press, 1992.

Kinnard, Douglas. *President Eisenhower and Strategy Management: A Study in Defense Politics.* Lexington: University Press of Kentucky, 1977.

———. *Secretary of Defense.* Lexington: University Press of Kentucky, 1980.

Kirkpatrick, Lyman B., Jr. *The Real CIA.* New York: Macmillan, 1968.

Kistiakowsky, George B. *A Scientist at the White House: The Private Diary of President Eisenhower's Special Assistant for Science and Technology.* Cambridge MA: Harvard University Press, 1976.

Klass, Philip J. *Secret Sentries in Space.* New York: Random House, 1971.

———. "The Truth about Missiles: What Went Wrong." *Aviation Week* 70 (December 1957): 35–64.

———. "USAF Probes Satellite Relay Stations." *Aviation Week* 69 (October 1958): 85–88.

Knapp, Wilfrid. *A History of War and Peace, 1939–65.* New York: Oxford University Press, 1967.

Koch, Scott A., ed. *Selected Estimates on the Soviet Union 1950–1959.* Washington DC: Center for the Study of Intelligence, CIA, 1993.

Koistinen, Paul A. C. *The Military-Industrial Complex: A Historical Perspective.* New York: Praeger, 1980.

Koppes, Clayton R. *JPL and the American Space Program: A History of the Jet Propulsion Laboratory.* New Haven CT: Yale University Press, 1982.

Krug, Linda T. *Presidential Perspectives on Space Exploration: Guiding Metaphors from Eisenhower to Bush.* New York: Praeger, 1991.

Kucera, Randolph P. *The Aerospace Industry and the Military: Structural and Political Relationships.* Beverly Hills CA: Sage, 1974.

Kulugin, Oleg Montuigne. *The First Directorate.* New York: St. Martin's Press, 1994.

LaFeber, Walter. *America, Russia, and the Cold War: 1945–1975.* New York: John Wiley & Sons, 1976.

———. *The American Age: United States Foreign Policy at Home and Abroad since 1750.* New York: Norton, 1989.

———. *Dynamics of World Power.* New York: Chelsea House, 1973.

Lang, Daniel. *From Hiroshima to the Moon.* New York: Simon & Schuster, 1959.

Larson, Arthur. *Eisenhower: The President Nobody Knew.* New York: Charles Scribner & Sons, 1968.

Lasby, Clarence G. *Project Paperclip: German Scientists and the Cold War.* New York: Atheneum, 1971.

Lattu, Kristan R., ed. *History of Rocketry and Astronautics.* American Astronautical Society Historical Series, vol. 8. San Diego: AAS, 1989.

"Launched Midas May Be First of Warning System." *Science News Letter,* June 4, 1960, 354–55.

Launius, Roger D. *NASA: A History of the U.S. Civil Space Program.* Malabar FL: Krieger, 1994.

———. *Organizing for the Use of Space: Historical Perspectives on a Persistent Issue.* AAS History Series, vol. 18. San Diego CA: American Astronautical Society, 1995.

Layton, Edwin T. *And I Was There.* New York: William Morrow, 1985.

Leary, William M. *The Central Intelligence Agency: History and Documents.* Tuscaloosa: University of Alabama Press, 1984.

Leffler, Melvin P. *A Preponderance of Power.* Stanford: Stanford University Press, 1992.

Lens, Sidney. *The Military-Industrial Complex.* Philadelphia: United Church Press, 1970.

Leslie, Stuart W. *The Cold War and American Science: The Military-Industrial-Academic Complex at MIT and Stanford.* New York: Columbia University Press, 1993.

Levantrosser, William F., ed. *Harry S. Truman: The Man from Independence.* Westport CT: Greenwood Press, 1986.

Levine, Alan J. *The Missile and Space Race.* Westport CT: Praeger, 1994.

Levine, Arthur L. *The Future of the US Space Program.* New York: Praeger, 1975.

Lewis, Jonathan E. *Spy Capitalism: ITEK and the CIA*. New Haven CT: Yale University Press, 2002.

Ley, Willy. *Rockets, Missiles, and Space Travel*. New York: Viking Press, 1961.

Licklider, Roy E. "The Missile Gap Controversy." *Political Science Quarterly* 85 (December 1970): 598–615.

"Lockheed Builds WS-117L Nerve Center." *Aviation Week* 71 (October 1959): 28.

Logsdon, John M. "Choosing Big Technology: Examples from the U.S. Space Program." *History and Technology* 9 (1992): 139–50.

Logsdon, John M., Dwayne A. Day, and Roger D. Launius, eds. *Exploring the Unknown: Selected Documents in the History of the US Civil Space Program*. Vol. 2: *External Relationships*. Washington DC: NASA, 1996.

Logsdon, John M., Linda J. Lear, Jannelle Warren-Findley, Ray A. Williamson, and Dwayne A. Day, eds. *Exploring the Unknown: Selected Documents in the History of the U.S. Civil Space Program*. Vol. 1: *Organizing for Exploration*. Washington DC: NASA, 1995.

Lowenthal, Mark M. *U.S. Intelligence: Evolution and Anatomy*. Westport CT: Praeger, 1992.

Ludwig, George H. *Opening Space Research: Dreams, Technology, and Scientific Discovery*. Google Books.

Lyon, Peter. *Eisenhower: Portrait of the Hero*. Boston: Little, Brown, 1974.

MacIsaac, David. *The Air Force and Strategic Thought 1945–1951*. Woodrow Wilson International Center for Scholars, International Security Studies Program, Working Paper No. 8. 1979.

Mack, Pamela E. "Making Big Technology Serve the User: U.S. Remote Sensing Programs." *History and Technology* 9 (1992): 95–107.

———. *Viewing the Earth: The Social Construction of the Landsat Satellite System*. Cambridge MA: MIT Press, 1990.

Maier, Charles S., ed. *The Origins of the Cold War and Contemporary Europe*. New York: New Viewpoints, 1978.

Marchetti, Victor, and John D. Marks. *The CIA and the Cult of Intelligence*. New York: Knopf, 1974.

Martz, Amy E. "The Organizational Culture of the National Aeronautics and Space Administration: A Historical Analysis." PhD dissertation, Pennsylvania State University, 1990. University Microfilm No. 91-17719.

Mason, Herbert Molly. *The United States Air Force: A Turbulent History*. New York: Mason/Charter, 1976.

May, Ernest, ed. *American Cold War Strategy: Interpreting NSC 68*. Boston: Bedford Books of St. Martin's Press, 1993.

Mazlish, Bruce. "Following the Sun." *Wilson Quarterly* 4 (Autumn 1980): 90–93.

———. *The Railroad and the Space Program: An Exploration in Historical Analogy*. Cambridge MA: MIT Press, 1965.

McAuliffe, Mary S. "Eisenhower, the President." *Journal of American History* 68 (December 1981): 624–32.

McCullough, David. *Truman*. New York: Simon & Schuster, 1992.

McDougall, Walter A. *. . . the Heavens and the Earth: A Political History of the Space Age*. New York: Basic Books, 1985.

———. "The Scramble for Space." *Wilson Quarterly* 4 (Autumn 1980): 71–82.

McElheny, Victor K. *Insisting on the Impossible: The Life of Edwin Land*. Reading MA: Perseus Books, 1998.

McNaugher, Thomas L. *Defense Management Reform: For Better or Worse*. Washington DC: Brookings Institution, 1990

———. *New Weapons, Old Politics: America's Military Procurement Muddle*. Washington DC: Brookings Institution, 1984.

McNeill, William H., *The Pursuit of Power: Technology, Armed Force and Society since 1000 A.D.* Chicago: University of Chicago Press, 1982.

McQuaid, Kim. *Big Business and Presidential Power: From FDR to Reagan*. New York: Morrow, 1982.

McQuain, Robert H. *Significant Achievements in Space Applications, 1966*. Washington DC: U.S. National Aeronautics and Space Administration, 1967.

Medaris, John B. *Countdown for Decision*. New York: G. P. Putnam's Sons, 1960.

Medvedev, Roy A., and Zhores A. Medvedev. *Khrushchev: The Years in Power*. Trans. Andrew R. Durkin. New York: Columbia University Press, 1976.

Melanson, Richard A., and David Mayers. *Reevaluating Eisenhower: American Foreign Relations in the 1950s*. Chicago: University of Illinois Press, 1987.

Melman, Seymour. *Decision-making and Productivity*. Oxford: Blackwell, 1958.

———. *The Defense Economy: Conversion of Industries and Occupations to Civilian Needs*. New York: Praeger, 1970.

———. *Pentagon Capitalism: The Political Economy of War*. New York: McGraw-Hill, 1970.

———. *The Permanent War Economy: American Capitalism in Decline*. New York: Simon & Schuster, 1974.

———. *The War Economy of the U.S.: Readings in Military Industry and Economy*. New York: St. Martin's Press, 1971.

Members of Congress for Peace through Law, Military Spending Committee. *The Economics of Defense: A Bipartisan Review of Military Spending*. New York: Praeger, 1971.

Metz, Steven. "Eisenhower and the Planning of American Grand Strategy." *Journal of Strategic Studies* 14 (March 1991): 49–71.

Meyers, William, and M. Vincent Hayes. *Conversion from War to Peace: Social, Economic, and Political Problems*. New York: Gordon and Breach, 1972.

Mezerik, A. G. *U-2 and Open Skies*. New York: International Review Series, 1960.

Miles, Carroll F. *The Office of the Secretary of Defense, 1947–1953*. New York: Garland, 1988.

Mills, C. Wright. *The Power Elite*. New York: Oxford University Press, 1959.

———. *Power Politics and People*. New York: Ballantine Books, 1963.

Moore, Otis C. "Twenty Years in Space: No Hiding Place in Space." *Air Force Magazine* 57 (1974): 43–48.

Moran, Theodore H. *American Economic Policy and National Security.* New York: Council of Foreign Relations Press, 1993.

Morenoff, Jerome. *World Peace through Space Law.* Charlottesville VA: Michie, 1967.

Morgan, Ian W. *Eisenhower versus "The Spenders": The Eisenhower Administration, the Democrats and the Budget, 1953–1960.* London: Pinter, 1990.

Moser, Don. "The Time of the Angel: The U-2, Cuba, and the CIA." *American Heritage* 28 (October 1977): 4–15.

Moskowitz, Charles C. *The Defense Sector and the American Economy.* New York: New York University Press, 1968.

Muenger, Elizabeth A. *Searching the Horizon: A History of Ames Research Center, 1940–1976.* Washington DC: NASA, 1985.

Mumford, Lewis. *The Myth of the Machine.* Vol. 2: *The Pentagon of Power.* New York: Harcourt, Brace & World, 1964.

———. *Technics and Civilization.* New York, Harcourt, Brace, 1963.

Murray, Bruce C. *Journey into Space: The First Three Decades of Space Exploration.* New York: Norton, 1989.

Nathan, James A. "A Fragile Detente: The U-2 Incident Re-examined." *Military Affairs* 39 (October 1975): 97–104.

National Science Foundation. *Research and Development in the Aircraft and Missile Industry, 1957–1968.* NSF 69-15. Washington DC: U.S. Government Printing Office, 1969.

Nelson, Anna Kasten. "The 'Top of Policy Hill': President Eisenhower and the National Security Council." *Diplomatic History* 7 (Fall 1983): 307–26.

Neufeld, Jacob. *The Development of Ballistic Missiles in the United States Air Force, 1945–1960.* Washington DC: Office of Air Force History, 1990.

———. *The Rocket and the Reich: Peenemunde and the Coming of the Ballistic Missile Era.* New York: Free Press, 1995.

Newell, Homer E., Jr. "International Geophysical Year Earth Satellite Program." *Scientific Monthly* 83 (July 1956): 13–21.

"New Space Tempo May Increase Support for WS-117L." *Aviation Week* 70 (May 1959): 26–27.

Oder, Frederic C. E., James C. Fitzpatrick, and Paul E. Worthman. "The CORONA Story." National Reconnaissance Office. November 1987.

Ordway, Frederick I., III, and Mitchell R. Sharpe. *The Rocket Team.* New York: Thomas Y. Crowell, 1979.

Pace, Steve. *Lockheed Skunk Works.* Osceola WI: Motor Books, 1992.

Parmet, Herbert S. *Eisenhower and the American Crusades.* New York: Macmillan, 1972.

Patterson, James T. *Mr. Republican: A Biography of Robert A. Taft.* Boston: Houghton Mifflin, 1972.

Payne, Rodger A. "Public Opinion and the Foreign Threats: Eisenhower's Response to Sputnik." *Armed Forces and Society* 21 (Fall 1994): 89–112.

Pearcy, Arthur. *Flying the Frontiers: NACA and NASA Experimental Aircraft.* Annapolis MD: Naval Institute Press, 1993.

Pearton, Maurice. *The Knowledgeable State: Diplomacy, War and Technology since 1830.* London: Burnett Books, 1982.

Peck, M. J., and F. M. Scherer. *The Weapons Acquisition Process: An Economic Analysis.* Boston: Harvard University Graduate School of Business Administration, 1962.

Peebles, Curtis. *The Corona Project: America's First Spy Satellites.* Annapolis MD: Naval Institute Press, 1997.

——. *Guardians: Strategic Reconnaissance Satellites.* Novato CA: Presidio Press, 1987.

——. *The Moby Dick Project: Reconnaissance Balloons over Russia.* Washington DC: Smithsonian Institution Press, 1991.

——. *Shadow Flights: America's Secret Air War against the Soviet Union.* Novato CA: Presidio Press, 2000.

Phillips, Almarin. *Technology and Structure: A Study of the Aircraft Industry.* Lexington MA: Heath Lexington Books, 1971.

Piore, Michael J., and Charles F. Sabel. *The Second Industrial Divide: Possibilities for Prosperity.* New York: Basic Books, 1984.

Possony, Stefan T. "Open Skies, Arms Control, and Peace." *Space Digest* 7 (March 1964): 71–72.

Prados, John. *The Soviet Estimate: US Intelligence Analysis and Soviet Strategic Forces.* Princeton NJ: Princeton University Press, 1986.

Prange, Gordon W. *At Dawn We Slept.* New York: Penguin Books, 1991.

"President Meets with Top Experts in Missile Review." *New York Times,* October 11, 1957.

"President Orders Missile Speed-up, Avoids Fund Rise." *New York Times,* October 13, 1957.

"President Pushes Rocket Program." *New York Times,* October 2, 1957.

Public Papers of the Presidents of the United States: Dwight D. Eisenhower 1953–1961. 8 vols. Washington DC: Government Printing Office, 1960–61.

Pursell, Carroll W., Jr. *The Military-Industrial Complex.* New York: Harper & Row, 1972.

Quester, George H. *Nuclear Diplomacy: The First 25 Years.* New York: Dunellen, 1970.

Rabe, Stephen G. "Eisenhower Revisionism: A Decade of Scholarship." *Diplomatic History* 17 (Winter 1993): 97–115.

Ramo, Simon. *Peacetime Uses of Outer Space.* New York: McGraw-Hill, 1961.

Ranelagh, John. *The Agency: The Rise and Decline of the CIA.* New York: Simon & Schuster, 1987.

Ransom, Harry Howe. *Central Intelligence and National Security.* Cambridge MA: Harvard University Press, 1958.

———. *The Intelligence Establishment.* Cambridge MA: Harvard University Press, 1970.

"Reds Win Race, Astound World." *Windsor (Ontario) Daily Star,* October 5, 1957.

Rhodes, Richard. "Analysis of the Cold War: The General and World War II." *New Yorker,* June 19, 1995, 47–59.

———. *Dark Sun: The Making of the Hydrogen Bomb.* New York: Simon & Schuster, 1995.

Rich, Ben R., and Leo Janos. *Skunk Works: A Personal Memoir of My Years at Lockheed.* New York: Little, Brown, 1994.

Richardson, Elmo. *The Presidency of Dwight David Eisenhower.* Lawrence: Regent Press of Kansas, 1979.

Richelson, Jeffrey. *American Espionage and the Soviet Target.* New York: William Morrow, 1987.

———. *America's Secret Eyes in Space.* New York: Harper & Row, 1990.

———. "The Keyhole Satellite Program." *Journal of Strategic Studies* (Great Britain) 7 (1984): 121–53.

———. *The U.S. Intelligence Community.* Boulder CO: Westview Press, 1999.

———. *The Wizards of Langley: Inside the CIA's Directorate of Science and Technology.* Boulder CO: Westview Press, 2001.

Ritchie, David. *Space War: The Fascinating and Alarming History of the Military Uses of Outer Space.* New York: Atheneum, 1982.

Roberts, C. M. *The Nuclear Years: The Arms Race and Arms Control, 1945–70.* New York: Praeger, 1970.

Roman, Peter J. "Eisenhower and Ballistic Missiles Arms Control, 1957–1960: A Missed Opportunity?" *Journal of Strategic Studies* 19 (September 1996): 365–80.

Rosen, Steven. *Testing the Theory of the Military-Industrial Complex.* Lexington MA: D. C. Heath, 1973.

Rosenberg, Nathan. *The Economics of Technological Change: Selected Readings.* Harmondsworth, England: Penguin, 1971.

———. *Inside the Black Box: Technology and Economics.* Cambridge MA: Cambridge University Press, 1982.

———. *Perspectives on Technology.* New York: Cambridge University Press, 1976.

———. *Technology and American Economic Growth.* New York: Harper & Row, 1972.

Rostow, W. W. *Open Skies: Eisenhower's Proposal of July 21, 1955.* Austin: University of Texas Press, 1982.

Ruffner, Kevin, ed. CORONA: *America's First Satellite Program.* Center for the Study of Intelligence. Washington DC: Government Printing Office, 1995.

———. "CORONA and the Intelligence Community." *Studies in Intelligence* 39 (1996): 61–69.

"Russia's 'Moon.'" *Wall Street Journal,* October 7, 1957.

Salkeld, Robert. *War and Space.* Englewood Cliffs NJ: Prentice-Hall, 1970.

Sarkesian, Sam C., ed. *The Military-Industrial Complex: A Reassessment.* Beverly Hills CA: Sage, 1972.

Sarkesian, Sam Charles, John Allen Williams, and Fred B. Bryant. *Soldiers, Society, and National Security.* Boulder CO: L. Rienner, 1995.

"Satellite Nears Final Design as Scientists Fight Deadline." *Aviation Week* 64 (January 1956): 37.

"Satellite Posing Legal Problems." *New York Times*, October 13, 1957.

Sayles, Leonard R. "James Webb at NASA." *Society* 29 (September/October 1992): 63–68.

Schaffel, Kenneth. *The Emerging Shield: The Air Force and the Evolution of Continental Air Defense, 1945–1960.* Washington DC: Office of Air Force History, 1991.

Schauer, William H. *The Politics of Space.* New York: Kolmer and Meir, 1976.

Schell, Jonathan. *The Time of Illusion.* New York: Knopf, 1976.

Scherer, Frederic M. *The Weapons Acquisition Process: Economic Incentives.* Boston: Division of Research, Graduate School of Business Administration, Harvard University, 1964.

Schilling, Warner R. "The H-Bomb Decision." *Political Science Quarterly* 76 (1961): 24–46.

Schlesinger, Arthur M., Jr. *The Dynamics of World Power: A Documentary History of U.S. Foreign Policy, 1945–1973.* 5 vols. New York: Chelsea House, 1973.

Schoettle, Enid Curtis Bok. "The Establishment of NASA." In *Knowledge and Power: Essays on Science and Government*, edited by Sanford A. Lakoff, 162–91. New York: Free Press, 1966.

Schultz, Kenneth W. "USAF's Conquest of Space." *Air Force Magazine* 57 (August 1974): 34–42.

Schwiebert, Ernest G. *A History of the U.S. Air Force Ballistic Missiles.* New York: Frederick A. Praeger, 1965.

"Sentry Satellite Shot Planned for Dec. 15." *Aviation Week* 69 (November 1958): 33.

Shaw, John E. "The Influence of Space Power upon History 1944–1998." *Air Power History* 46 (Winter 1999): 20–29.

Sherry, Michael S. *In the Shadow of War: The US since the 1930s.* New Haven CT: Yale University Press, 1995.

———. *The Rise of American Air Power: The Creation of Armageddon.* New Haven CT: Yale University Press, 1987.

Shternfeld, Ari. *Soviet Space Science.* New York: Basic Books, 1959.

Shulsky, Abram N. *Silent Warfare: Understanding the World of Intelligence.* Washington DC: Brassey's, 1993.

Sibley, Katherine A. S. "Soviet Industrial Espionage against American Military Technology and the U.S. Response, 1930–1945." *Intelligence and National Security* 14 (Summer 1999): 94–123.

Sigethy, Robert. "The Air Force Organization for Basic Research, 1945–70." PhD dissertation, American University, 1988. DAI 41:4491.

Simonsen, G. R., ed. *The History of the American Aircraft Industry: An Anthology.* Cambridge MA: MIT Press, 1968.

Slack, Michael, and Heather Chestnutt. *Open Skies: Technical, Organizational, Operational, Legal and Political Aspects.* Toronto: York University, Centre for International and Strategic Studies, 1990.

Slesinger, Reuben E. *National Economic Policy: The Presidential Reports.* Princeton NJ: D. Van Nostrand, 1968.

Sloop, John L. *History of Rocketry and Astronautics: Proceedings of the Seventeenth History Symposium of the International Academy of Astronautics.* AAS Historical Series, vol. 12. IAA History Symposium, vol. 7. San Diego CA: AAS, 1991.

Smith, Bruce L. R. *The RAND Corporation: Case Study of a Nonprofit Advisory Corporation.* Cambridge MA: Harvard University Press, 1966.

Smith, Marcia S. "America's International Space Activities." *Society* 21 (January–February 1984): 18–25.

Smoke, Richard. *National Security and the Nuclear Dilemma: An Introduction to the American Experience.* Reading MA: Addison-Wesley, 1984.

Soapes, Thomas F. "A Cold Warrior Seeks Peace: Eisenhower's Strategy for Nuclear Disarmament." *Diplomatic History* 4 (Winter 1980): 57–71.

"Soviet Fires Earth Satellite into Space; It Is Circling the Globe at 18,000 M.P.H.; Sphere Traced in 4 Crossings over U.S." *New York Times,* October 5, 1957.

"Soviets Launch Space Era." *Windsor (Ontario) Daily Star,* October 5, 1957.

"Soviets Play on Propaganda." *New York Times,* October 12, 1957.

"Soviets Predict Their Earth Satellite Will Force U.S. Foreign Policy Change: 'Moonshine,' Says U.S. Aide." *Wall Street Journal,* October 8, 1957.

"Space Law: A Symposium." Prepared at the Request of Honourable Lyndon B. Johnson, Chairman. Special Committee on Space and Astronautics. U.S. Senate, 85th Congress, 2nd Session. December 31, 1958. Washington DC: Government Printing Office, 1959.

Stares, Paul B. *Space and National Security.* Washington DC: Brookings Institute, 1987.

——. "Space and U.S. National Security." *Journal of Strategic Studies* 6 (December 1983): 31–48.

Stassen, Harold, and Marshall Houts. *Eisenhower: Turning the World towards Peace.* St Paul MN: Merrill/Magnus, 1990.

Steel, Ronald. *Pax Americana.* Rev. ed. New York: Viking, 1970.

Steinberg, Gerald M. *Satellite Reconnaissance: The Role of Informal Bargaining.* New York: Praeger, 1983.

Steury, Donald, ed. *Estimates on Soviet Military Power, 1954–1984.* Washington DC: Center for the Study of Intelligence, 1996.

——, ed. *Intentions and Capabilities: Estimates on Soviet Strategic Forces, 1950–1983.* Washington DC: Government Printing Office, 1996.

——, ed. *Sherman Kent and the Board of National Estimates: Collected Essays.* Washington DC: History Staff, Center for the Study of Intelligence, CIA, 1994.

Stone, Irving. "U.S. Schedules Additional Samos Launches." *Aviation Week* 73 (October 1960): 28–29.

Stuhlinger, Ernst, and Frederick Ira Ordway. *Wernher von Braun, Crusader for Space: A Biographical Memoir.* Malabar FL: Krieger, 1996.

Sudoplatov, Pavel, Anatoli Sudoplatov, Jerrold L. Schecter, and Leana P. Schecter. *Special Tasks.* New York: Little, Brown, 1994.

Sullivan, Walter. *Assault on the Unknown: The International Geophysical Year.* New York: McGraw-Hill, 1961.

Suri, Jeremi. "America's Search for a Technological Solution to the Arms Race: The Surprise Attack Conference of 1958 and a Challenge for 'Eisenhower Revisionists.'" *Diplomatic History* 21 (Summer 1997): 417–51.

Sweeney, Richard. "Discoverer II Orbital Attitude Controlled." *Aviation Week* 70 (April 1959): 26–27.

Swomley, John M., Jr. *The Military Establishment.* Boston: Beacon Press, 1946.

Taubman, Philip. *Secret Empire: Eisenhower, the CIA, and the Hidden Story of America's Space Espionage.* New York: Simon & Schuster, 2003.

Taylor, John, and David Monday. *Spies in the Sky.* New York: Scribner, 1972.

Terry, Michael R. "Formulation of Aerospace Doctrine from 1955 to 1959." *Air Power History* 38 (Spring 1991): 47–54.

"Truth about Missiles: Excerpts from Testimonies before Senate Committee." *U.S. News and World Report* 43 (December 1957): 35–38.

Tyler, C. Merton. *Pentagon Partner: The New Nobility.* New York: Grossman, 1970.

Udis, Bernard, ed. *The Economic Consequences of Reduced Military Spending.* Lexington MA: D. C. Heath, 1973.

Ujvarosy, Kazmer. "Aircraft and Spacecraft in U.S. Intelligence Activities: Reconnaissance, Surveillance, and Special Operations." MA thesis, San Francisco State University, 1985.

"USAF Pushes Pied Piper Space Vehicle." *Aviation Week* 67 (October 1957): 26.

"USAF Strengthens Samos Effort." *Aviation Week* 73 (September 12, 1960): 31.

U.S. Assistant Secretary of Defense (Comptroller). *The Economics of Defense Spending: A Look at the Realities.* Washington DC: Government Printing Office, 1972.

U.S. Congressional Record. 82nd Congress, 1st–2nd Sessions. Reel 323, vol. 98, pt. 1-2. 1-2786.

U.S. Department of State. *Documents on Disarmament 1945–1959.* Vol. 1. Washington DC: Government Printing Office, 1960.

———. *Foreign Relations of the United States 1945–1950: Emergence of the Intelligence Establishment.* Washington DC: Government Printing Office, 1996.

———. *Foreign Relations of the United States 1952–1954: General: Economic and Political Matters.* Vol. 1, pt. 1. Washington DC: Government Printing Office, 1983.

———. *Foreign Relations of the United States 1952–1954: General: Economic and Political Matters.* Vol. 1, pt. 2. Washington DC: Government Printing Office, 1983.

———. *Foreign Relations of the United States 1952–1954: National Security Affairs.* Vol. 2, pt. 1. Washington DC: Government Printing Office, 1980.

———. *Foreign Relations of the United States 1952–1954: National Security Affairs.* Vol. 2, pt. 2. Washington DC: Government Printing Office, 1984.

———. *Foreign Relations of the United States 1952–1954: The Geneva Conference.* Vol. 16. Washington DC: Government Printing Office, 1981.

———. *Foreign Relations of the United States 1955–1957: Foreign Aid and Economic Defense.* Vol. 10. Washington DC: Government Printing Office, 1989.

———. *Foreign Relations of the United States 1955–1957: Foreign Economic Policy; Foreign Information Program.* Vol. 9. Washington DC: Government Printing Office, 1987.

———. *Foreign Relations of the Untied States 1955–1957: National Security Policy.* Vol. 19. Washington DC: Government Printing Office, 1990.

———. *Foreign Relations of the United States 1955–1957: Soviet Union; Eastern Mediterranean.* Vol. 24. Washington DC: Government Printing Office, 1989.

———. *Foreign Relations of the United States, 1955–1957: United Nations and General International Matters.* Vol. 11. Washington DC: Government Printing Office, 1988.

"U.S. Studies Missile Speed Up." *Windsor (Ontario) Daily Star,* October 10, 1957.

Van Allen, James A. "The Artificial Satellite as a Research Instrument." *Scientific American* 195 (November 1956): 41–47.

Vander Meulen, Jacob A. *The Politics of Aircraft: Building an American Military Industry.* Lawrence: University Press of Kansas, 1991.

Vexler, Robert I., ed. *Dwight D. Eisenhower 1890–1969: Chronology, Documents and Bibliographical Aids.* Dobbs Ferry NY: Oceana, 1970.

Von Braun, Werner, and Frederick I. Ordway III. *History of Rocketry and Space Travel.* New York: Crowell, 1975.

Von Braun, Werner, Frederick I. Ordway III, and David Dooling. *Space Travel: A History.* New York: Harper & Row, 1985.

Warshaw, Shirley Ann. *Reexamining the Eisenhower Presidency.* Westport CT: Greenwood Press, 1993.

"Watch from Space." *Newsweek* 55 (January 1960): 39–40.

Watson, George M., Jr. *The Office of the Secretary of the Air Force, 1947–1965.* Washington DC: Center for Air Force History, 1993.

Watson, Robert J. *The History of the Joint Chiefs of Staff.* Vol. 5: *The Joint Chiefs of Staff and National Policy 1953–1954.* Washington DC: Historical Division Joint Chiefs of Staff, 1986.

Weidenbaum, Murray L. *The Economics of Peacetime Defense.* New York: Praeger, 1975.

Wells, William G., Jr. "Science Advice and the Presidency, 1933–1976." PhD dissertation, George Washington University, 1977. DAI 39:1832-A.

Welzenbach, Donald E. "Din Land: Patriot from Polaroid." *Optics and Photonics News* 5 (October 1994): 22–29.

Wilson, Andrew. *The Eagle Has Wings: The Story of American Space Exploration, 1945–1975.* London: British Interplanetary Society, 1982.

Winter, Frank H. *Rockets into Space*. Cambridge MA: Harvard University Press, 1990.

Wohlstetter, Roberta. *Pearl Harbor: Warning and Decision*. Stanford: Stanford University Press, 1962.

Yaffee, Michael. "First ARPA Satellite Launches to Seek Key to Improved Vehicle." *Aviation Week* 69 (December 1958): 32.

Yarmolinsky, Adam. *The Military Establishment: Its Impact on American Society*. New York: Harper & Row, 1971.

———. *United States Military Power and Foreign Policy*. Chicago: University of Chicago Center for Policy Study, 1967.

York, Herbert. *The Advisors: Oppenheimer, Teller and the Super Bomb*. San Francisco: W. H. Freeman, 1976.

———. *Arms and the Physicist*. Woodbury NY: American Institute of Physics, 1995.

———. *Race to Oblivion: A Participant's View of the Arms Race*. New York: Simon & Schuster, 1970.

"York Outlines Satellite Plans." *Aviation Week* 69 (September 1958): 19.

Zaloga, Steven J. *Target America: The Soviet Union and the Strategic Arms Race, 1945–1964*. Novato CA: Presidio, 1993.

Ziegler, Charles A., and David Jacobson. *Spying without Spies: Origins of America's Secret Nuclear Surveillance System*. Westport CT: Praeger, 1995.

INDEX

Gardner, Trevor, 60, 66
Gates, Thomas S., Jr., 177
General Electric, 87
General Operational Requirement No. 80, 125, 231n27
GENETRIX program (WS-119L), 13–14, 206n36
German scientists in Soviet Union, 8, 19, 86
Ginsburg, Charles, 123
Glasser, Otto J., 128
Goddard, George W., 96
Goldmark, Peter, 130
Goodpaster, Andrew J., 36, 67, 133, 157, 177, 215n65
Goodwin, Craufurd, 40
GOPHER, 99, 106–7
Gray, Gordon, 177
Greene, Sidney, 116, 118, 122
Greenstein, Fred: "Eisenhower as an Activist President," 191–92
Greer, Kenneth E.: "CORONA," 197
Groves, Leslie, 9

Hagerty, Jim, 78
Hall, Edward, 119
Hall, Harvey, 86
Hall, R. Cargill, 25, 68, 87, 90, 109, 116, 125; "Origins of U.S. Space Policy," 198
Hanscom Air Force Base, 124
Harris, Seymour, 40
Haydon, Brownlee W., 142
Hazlett, Everett "Swede," 31–32, 55, 56
The Heavens and the Earth (McDougall), 196
HERMES, 87
Herther, John C., 125–26, 128
high-altitude aircraft. See U-2 aircraft
high-altitude balloons, 105, 106–7
historiography of satellite reconnaissance development: CIA and, 189, 193–95; CORONA and, 189; Eisenhower and, 190–93; post-declassification, 196–98; pre-declassification, 195–96; and scarcity of information on WS-117L, xiii–xiv, 189–90, 195
"History of Satellite Reconnaissance" (Perry), 130–31

Holding the Line (Alexander), 190
horizontal scanner for WS-117L, 118–19
hot-air balloons, reconnaissance and, 99
Huckaby, Jim, 119, 122–23
Humphrey, George, 46, 79, 82
Huntington, Samuel, 53
Huntzicker, J. H., 145; "Physical Recovery of Satellite Payloads," 142–43
HUSTLER. See B-58 HUSTLER
"Hyac" (high acuity) camera, 145

IBM, 127
Ike's Spies (Ambrose), 192
Image Orthicon television camera, 112
inflation, 31, 40–42, 43
intelligence gathering: after Sputnik, 146; Edwin Land's 1954 panel on, 62–66, 81, 177, 178; Eisenhower administration and, 15–25; Pearl Harbor attack's influence on, 4–6, 15, 24, 28, 47, 95–98; Truman administration and, 3–15
intercontinental ballistic missiles (ICBMs): expectations of Soviet development of, 12, 18–19; hydrogen bomb and repercussions for, 23–24; as means to deliver nuclear weapons, 14–15; military budget cuts and, 55; post-Sputnik missile gap and, 138; propulsion for satellites and, 66; reentry technology and, 142, 143; satellite planning and, 101; separation of, from civilian space activities, 78–79
intermediate range ballistic missiles (IRBMs), 152. See also THOR IRBM program
International Geophysical Year (IGY), xi, 59, 72–76, 82, 115, 131. See also VANGUARD
international law in space: after Sputnik, 80, 146; and freedom of space, 65–66, 73, 75, 76, 82; Open Skies initiative and, 59, 62, 69–72, 80, 81, 115, 221n35; secret planning for U-2 and, 67–69
interservice rivalry. See military service rivalries
"Invitation to Struggle" (Day), 198
Itek Corporation, 106, 235n24

Wilson, Charles, 52, 54, 67, 79
Wisner, Frank G., 36
Wohlstetter, Albert, 21
Wright Air Development Center (WADC),
 WS-117L and, 108, 116, 117–26; and fund-
 ing and contractor issues, 117–19, 120–22,
 124–26; and technical issues, 118, 119–20,
 122–24; Western Development Division
 stage and, 127–28
Wright Field, 92–93, 96, 97, 103–4, 126
WS-117L satellite program, 3, 74, 82, 108;
 ARPA and, 164–65, 181; complexity of,
 163–64; contractors and, 118–19, 121–22,
 126, 127–30, 132; declassification and,
 197–99; electronics and, 119–20; and
 Program II-A, 152, 153, 154–60, 181; and
 Programs IV, VI, and VIII, 152–53; scarcity of information on, xiii–xiv, 115–16,
135–36, 140, 141, 189–90, 195; slow progress and, 126, 147; *Sputnik*'s impact
on, 137, 145–49, 186; theoretical barriers broken by, 186; as three satellites,
140–41, 161, 165–66, 182–83; as USAF program, xiv, 189; various designations of,
108, 116–17, 156; Western Development
Division stage of, 126–35; Wright Air
Development Center stage of, 116, 117–
26. *See also* CORONA; DISCOVERER;
MIDAS (Missile Defense Alarm System);
SENTRY/SAMOS

"year of maximum danger" concept, 30–
31, 42–43
York, Herbert, 114, 162, 164, 170, 175–77,
242n41

03/16